ASCE Manuals and Reports on Engineering Practice No. 139

Radar Rainfall Data Estimation and Use

Prepared by the
Radar Rainfall Data and Application Task Committee of the
Surface Water Hydrology Technical Committee of the
Watershed Council of the Environmental and Water Resources
Institute of the American Society of Civil Engineers

Edited by
Chandra S. Pathak, Ph.D., P.E.
and Ramesh S. V. Teegavarapu, Ph.D., P.E.

ASCE AMERICAN SOCIETY OF CIVIL ENGINEERS

EWRI ASCE
ENVIRONMENTAL & WATER RESOURCES INSTITUTE

Published by the American Society of Civil Engineers

Library of Congress Cataloging-in-Publication Data

Names: Pathak, Chandra S., editor. | Teegavarapu, Ramesh S. V., 1970- editor. | Environmental and Water Resources Institute (U.S.). Radar Rainfall Data and Application Task Committee.
Title: Radar rainfall data: estimation and use / prepared by the Radar Rainfall Data and Application Task Committee, of the Surface Water Hydrology Technical Committee, of the Watershed Council, of the Environmental and Water Resources Institute, of the American Society of Civil Engineers; edited by Chandra S. Pathak, Ph.D., P.E., D.WRE [and] Ramesh S.V. Teegavarapu, P.E.
Other titles: Rainfall data
Description: Reston, Virginia: American Society of Civil Engineers, [2018] | Series: ASCE manuals and reports on engineering practice; no. 139 | Includes bibliographical references and index.
Identifiers: LCCN 2018029381 | ISBN 9780784415115 (hardcover: alk. paper)
Subjects: LCSH: Radar meteorology. | Rain and rainfall—Forecasting. | Rain and rainfall.
Classification: LCC QC973.5 .R325 2018 | DDC 551.63/53—dc23
LC record available at https://lccn.loc.gov/2018029381

Published by American Society of Civil Engineers
1801 Alexander Bell Drive
Reston, Virginia 20191-4382
www.asce.org/bookstore | ascelibrary.org

***Errata:* Errata, if any, can be found at https://doi.org/10.1061/9780784415115.**

ISBN 978-0-784-1511-5 (print)
ISBN 978-0-7844-8176-9 (PDF)
ISBN 978-0-7844-8177-6 (ePub)
Manufactured in the United States of America.

24 23 22 21 20 19 18 1 2 3 4 5

Photo credits: — front cover, radar tower: https://www.flickr.com/photos/noaaphotolib/9672381030/in/album-72157638866298193/

— front cover, map and visual of radar data: https://www.flickr.com/photos/noaaphotolib/5278185262/in/photolist-93q5Ab-93mX7K-8QvfYj-8QsbPc

— back cover, weather map: https://www.flickr.com/photos/noaaphotolib/9627283271/in/album-72157625168461902/

MANUALS AND REPORTS ON ENGINEERING PRACTICE

(As developed by the ASCE Technical Procedures Committee, July 1930, and revised March 1935, February 1962, and April 1982)

A manual or report in this series consists of an orderly presentation of facts on a particular subject, supplemented by an analysis of limitations and applications of these facts. It contains information useful to the average engineer in his or her everyday work, rather than findings that may be useful only occasionally or rarely. It is not in any sense a "standard," however; nor is it so elementary or so conclusive as to provide a "rule of thumb" for nonengineers.

Furthermore, material in this series, in distinction from a paper (which expresses only one person's observations or opinions), is the work of a committee or group selected to assemble and express information on a specific topic. As often as practicable the committee is under the direction of one or more of the Technical Divisions and Councils, and the product evolved has been subjected to review by the Executive Committee of the Division or Council. As a step in the process of this review, proposed manuscripts are often brought before the members of the Technical Divisions and Councils for comment, which may serve as the basis for improvement. When published, each work shows the names of the committees by which it was compiled and indicates clearly the several processes through which it has passed in review, so that its merit may be definitely understood.

In February 1962 (and revised in April 1982), the Board of Direction voted to establish a series titled "Manuals and Reports on Engineering Practice," to include the Manuals published and authorized to date, future Manuals of Professional Practice, and Reports on Engineering Practice. All such Manual or Report material of the Society would have been refereed in a manner approved by the Board Committee on Publications and would be bound, with applicable discussion, in books similar to past Manuals. Numbering would be consecutive and would be a continuation of present Manual numbers. In some cases of joint committee reports, bypassing of Journal publications may be authorized.

A list of available Manuals of Practice can be found at http://www.asce.org/bookstore.

CONTENTS

PREFACE ix

ACKNOWLEDGMENTS xi

1. RADAR RAINFALL ESTIMATION 1
   1.1 Introduction 1
   1.2 Background 2
   1.3 Scope 7
   1.4 Availability of Radar Rainfall Data within the United States ... 7
   References 8

2. RADAR RAINFALL DATA: TEMPORAL AND SPATIAL CHARACTERISTICS 11
   2.1 Native Radar Data Resolution 11
   2.2 Radar Rainfall Data Mosaics 12
   2.3 Data Formats and Resolutions 13
   2.4 Radar Rainfall Data QA/QC and Data Management 16
   2.5 Gauge-Adjusted Radar Rainfall Estimates 20
   2.6 Tool for Radar Rainfall Data Analysis (HEC-MetVue) 21
   2.7 Use of Radar Rainfall Data 22
   2.8 Radar Rainfall Data Issues and Future Perspectives 26
   2.9 Conclusions 27
   References 28

3. RADAR RAINFALL DATA PROCESSING 31
   3.1 Background 31
   3.2 Data Acquisition and Processing 33
   3.3 Reflectivity–Precipitation Rate Relationships 38
   3.4 Radar QPE Products from the WSR-88D Radar Product Generator 40

3.5 Error Distribution of Radar Rainfall Estimates ... 40
3.6 Approaches to Gauge–Radar Adjustment ... 41
3.7 Approaches to Gauge–Radar Observation Merging ... 43
3.8 Applicability of the Gauge–Radar Approaches ... 45
3.9 Use of Daily Precipitation Reports in Combination with Radar QPE ... 47
3.10 Access to Precipitation Observations and Estimates ... 48
3.11 Conclusions ... 49
References ... 49

4. EVALUATION AND IMPROVEMENT OF RADAR RAINFALL DATA ... 53
4.1 Rainfall Measurement Methods, Errors, and Accuracy ... 53
4.2 Rain-Gauge and Radar-Based Measurements ... 53
4.3 Improving Radar-Based Estimation: Optimal *Z–R* Relationships ... 54
4.4 Conclusions ... 57
References ... 57

5. USE OF RADAR RAINFALL DATA IN HYDROLOGIC MODELING ... 59
5.1 Data Requirements for Hydrologic Modeling and Design ... 59
5.2 Radar-Based Rainfall Data for Hydrologic Modeling ... 60
5.3 Conclusions ... 62
References ... 62

6. EXAMPLES IN RADAR RAINFALL DATA, ANALYSES, AND APPLICATIONS ... 63
6.1 Radar Rainfall Estimation—South Florida Water Management District ... 63
6.2 Radar Rainfall Data Analyses ... 65
6.3 Other Radar Rainfall Data Applications ... 66
6.4 Conclusions ... 69
References ... 69

7. ADVANCED TOPIC: FRAMEWORK FOR BIAS ANALYSIS OF RADAR DATA ... 73
7.1 Bias Analysis Methods ... 73
7.2 Ideal Performance Measures and Skill Scores ... 84
7.3 Utility of Assessment Indexes and Performance Measures ... 85
7.4 Bias Corrections ... 89
7.5 Bias Corrections with Limited Rain-Gauge Data ... 90
7.6 Bias Corrections: Temporal Resolution Issues ... 90
7.7 Conclusions ... 91
References ... 91

8. ADVANCED TOPIC: RAIN-GAUGE RAINFALL DATA AUGMENTATION AND RADAR RAINFALL DATA ANALYSIS 95
8.1 Spatial and Temporal Analysis of Rainfall 95
8.2 Missing Data Estimation 96
8.3 Use of Radar Data for Infilling Rainfall Data 98
8.4 Geospatial Grid-Based Transformations of Radar-Based Rainfall Data 102
8.5 Issues with Filled Precipitation Data Series 105
8.6 Conclusions 107
References 107

9. ADVANCED TOPIC: DESIGN OF RAINFALL MONITORING NETWORKS 111
9.1 Design of Rainfall Monitoring Networks 111
9.2 Rain-Gauge Network Density 111
9.3 Optimal Rain-Gauge Monitoring Networks 112
9.4 Optimal Density and Monitoring Networks 113
9.5 Objectives for Monitoring Network Design 113
9.6 Optimal Monitoring Network Design 114
9.6 Optimal Network Design Using Radar Data 116
9.7 Post-Network Design Recommendations for Rain-Gauge Placements 118
9.8 Identification of Meteorologically Homogeneous Areas 118
9.9 Conclusions 119
References 120

INDEX 121

PREFACE

This manual on radar rainfall data estimation was developed as one of the several tasks that was undertaken by the Radar Rainfall Data and Application Task Committee, under the Surface Water Hydrology Technical Committee within the Watershed Council of Environmental and Water Resources Institute's (EWRI), American Society of Civil Engineers (ASCE).

This manual was developed primarily for use by the practicing water resources engineers in the industry. This manual has nine chapters and their titles and authors are shown as follows.

Chapter 1: Radar Rainfall Estimation, Chandra S. Pathak (US Army Corps of Engineers);

Chapter 2: Radar Rainfall Data: Temporal and Spatial Characteristics, Chandra S. Pathak (US Army Corps of Engineers) and David C. Curtis (WEST Consultants);

Chapter 3: Radar Rainfall Data Processing, David Kitzmiller (National Weather Service);

Chapter 4: Evaluation and Improvement of Radar Rainfall Data, Ramesh S.V. Teegavarapu (Florida Atlantic University);

Chapter 5: Use of Radar Rainfall Data in Hydrologic Modeling, David C. Curtis (WEST Consultants);

Chapter 6: Examples in Radar Rainfall Data, Analyses, and Applications, Chandra S. Pathak (US Army Corps of Engineers);

Chapter 7: Advanced Topic: Framework for Bias Analysis of Radar Data, Ramesh S.V. Teegavarapu (Florida Atlantic University);

Chapter 8: Advanced Topic: Rain-Gauge Rainfall Data Augmentation and Radar Rainfall Data Analysis, Ramesh S.V. Teegavarapu (Florida Atlantic University);

Chapter 9: Advanced Topic: Design of Rainfall Monitoring Networks, Ramesh S.V. Teegavarapu (Florida Atlantic University).

ACKNOWLEDGMENTS

The Blue-Ribbon Panel of reviewers was established by the Watershed Council of EWRI and approved by the Technical Executive Committee of EWRI. The Blue-Ribbon Panel Review members were Dr. Scott Huebner, P.E. (South Florida Water Management District); Mr. Michael DelCharco, P.E. (Taylor Engineering, Inc.); and Mr. David Preusch, P.E. (WEST Consultants, Inc.). The drafts of the manual were reviewed by individuals chosen for their diverse perspective and technical expertise in accordance with the procedures approved for ASCE's *Manual of Practice*. The purpose of this independent review was to provide candid and critical comments that assist EWRI in making this publication as sound as possible and to ensure that the manual meets ASCE institutional standards in science and engineering and usability by the practicing engineers.

The authors wish to thank the following individuals for their reviews of this manual: Dr. Scott Huebner, P.E. (South Florida Water Management District); Mr. Michael DelCharco, P.E. (Taylor Engineering, Inc.); Mr. David Preusch, P.E. (WSP USA and formerly with WEST Consultants, Inc.); Dr. Thomas Evans (US Army Corps of Engineers); and Dr. Daniel Wright (University of Wisconsin—Madison and formerly with Princeton University).

CHAPTER 1
RADAR RAINFALL ESTIMATION

1.1 INTRODUCTION

Radar-derived rainfall estimation is clearly one of the most significant recent advances in hydrologic engineering and practice. There is a trade-off in hydrologic modeling between using rainfall data collected by rain gauges and those collected by radar. Rain gauges provide point values of rainfall depth and intensity, but they are not cost-effective in providing information about the spatial distribution of rainfall. Although rain gauges might suffice for widespread rain events, a gauge network can miss localized convective rainfall events altogether. Radar-derived rainfall data provide a density of measurements that are not obtainable by rain gauges alone. Combining these two sensor systems produces better rainfall estimates that more accurately characterize rainfall over a watershed. Using rain-gauge–adjusted radar rainfall data, hydrologists have more information about rainfall rates at high resolution in space and time over large areas. Understanding radar rainfall estimation and estimation errors within a hydrologic modeling context have important advantages.

The *Journal of Hydrologic Engineering* has been publishing technical articles on radar rainfall since 1996. Through April 2013, it has published more than 150 technical papers, case studies, and technical notes on this subject. In February 2013, the *Journal of Hydrologic Engineering* published a special issue on "Radar Rainfall Data Analyses and Applications." This publication includes the most significant advances in hydrologic engineering practice on the subject. This special issue (Pathak 2013) has 17 technical papers grouped into four categories: (a) two papers are of radar rainfall data development methods, (b) six papers are on radar rainfall data improvement and validation, (c) four papers are on application of radar rainfall data, and (d) five papers are on case studies. The reader is also directed to the special issue on radar rainfall and operational hydrology (Pathak et al. 2017) and a special issue on hydrologic applications of weather radar (Seo et al. 2015) for a compendium of papers on different issues ranging from radar estimation to applications for operational hydrology.

This manual is intended to provide a flexible framework within which both government agencies and private consulting firms can develop radar rainfall data sets and analyze them according to their varied goals and resources. The guidelines herein are not intended to provide a complete manual of all procedures available for estimating radar rainfall data and data analyses. The basic philosophy and principles are described in sufficient detail to promote a reasonable degree of consistency and uniformity among data developers and users to perform various tasks within the hydrologic engineering field.

We, the editors and the authors, sincerely hope that this manual on the topic of radar rainfall data analyses and applications is useful to researchers and practicing engineers alike as this discipline of hydrologic engineering progresses in the future. This manual was developed as a part of several tasks that were undertaken by the Radar Rainfall Data and Application Task Committee, under the Surface Water Hydrology Technical Committee within the Watershed Council of the Environmental and Water Resources Institute (EWRI) of ASCE.

1.2 BACKGROUND

1.2.1 Traditional Rainfall Estimation

Traditionally, rainfall has been and is measured by rain gauges—devices that record rainfall entering the gauge orifice. Gauge diameters generally range from 1 to 12 in. (2.54 to 30.48 cm) in diameter. Even the largest diameter gauge (12 in. or 30.48 cm) measures rainfall over only a tiny portion of a watershed (0.0000000282 mi^2 or 0.0000000730 km^2). With a relatively dense network density of one gauge per 10 mi^2 (26 km^2), rainfall is only measured in parts per billion.

Hydrologically, rainfall estimates at tiny points are not particularly valuable. By themselves, rain-gauge measurements represent very little volume. What is really needed are rainfall estimates over an entire watershed to accurately determine watershed volumetric inflow. To estimate rainfall over an area from point rain-gauge data, engineers and hydrologists use a variety of interpolation techniques to compute a rainfall "surface" over a watershed. The rainfall surface is then integrated over the watershed area during each time step to estimate the volumetric rate of inflow (Figure 1-1).

Figure 1-1. Example of an automated rain gauge.

Common rainfall interpolation techniques include Thiessen polygons (Viessman et al. 1972), inverse distance squared weighting (Chow et al. 1988), and kriging (Bras and Rodríquez-Iturbe 1985). Each interpolated method is some form of weighted averaging of individual rain-gauge observations to create an area averaged rainfall estimate.

Figure 1-2 shows a rainfall "surface" interpolated from rainfall observed at gauge locations using the Thiessen polygon technique. A polygon enclosing all points closest to a given rain gauge is a Thiessen polygon. Each point within a Thiessen polygon is assigned the same value as the enclosed rain-gauge observation. By comparison, Figure 1-3 shows a rainfall surface for the same event as represented by radar rainfall estimates.

As seen in Figure 1-3, the band of heavy rainfall to the east is missed by several of the rain gauges. As a result, the Thiessen polygon estimates underrepresent rainfall in the eastern portion of the watershed (the watershed boundary is shown in white) and overrepresent rainfall in the middle portion of the watershed.

1.2.2 Radar Rainfall Estimation

The use of radar to estimate rainfall began in the 1960s. During the early 1990s, the use of this technology proliferated as the NOAA (National Oceanic and Atmospheric Administration) installed radar stations across the United States as part of the Weather Surveillance Radar 88-Doppler (WSR-88D) and the NEXt Generation RADar (NEXRAD) program initiated by the NWS (National Weather Service). Currently, 160 WSR-88D radar

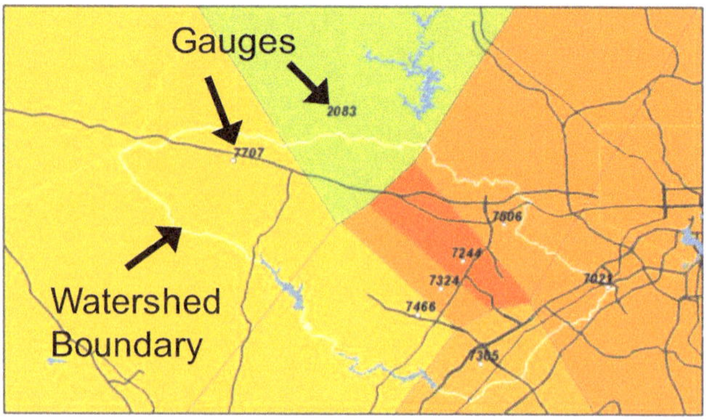

Figure 1-2. Thiessen polygon representation of rainfall.

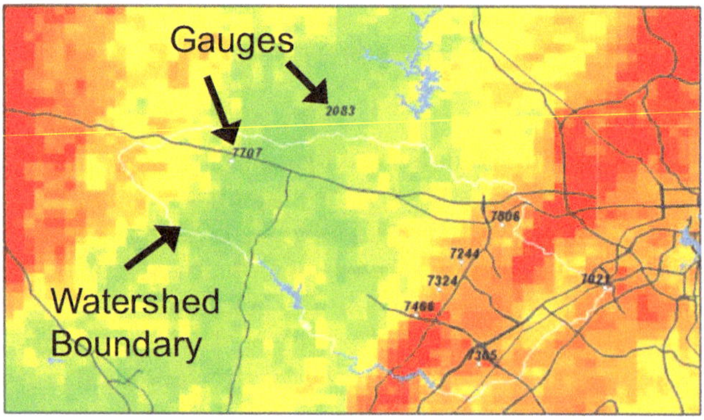

Figure 1-3. Gauge-adjusted radar rainfall.

stations are in operation as part of the NEXRAD network, providing estimates of rainfall and snowfall intensity with nearly complete spatial coverage across most of the United States.

Doppler (WSR-88D) radar operates by emitting short pulses of microwave energy. When a target such as a building, airplane, bird, snowflake, or rain droplet is encountered, the emitted energy is scattered in all directions. Small amounts of energy, known as backscatter, are returned to the radar antenna, where they are detected and recorded, known as radar reflectivity. The intensity of the returned radar reflectivity signal is then related to the size of the object and analyzed according to the time required for the pulse to reach the target and return to the antenna. This provides information regarding the range and Doppler velocity of the target relative to the radar. The WSR-88D radar is a 10 cm-wavelength (S-band) radar. It is designed for long-distance surveillance because its wavelength penetrates rainfall with little reduction in the strength of the signal (minimal attenuation). However, European radar systems predominantly use shorter- (less than 5 cm) wavelength (C-band) radar. Shorter-wavelength radars have relatively greater attenuation caused by absorption and scattering of the electromagnetic radiation that degrades performance as the distance from the radar increases, thus requiring an increased density of radar systems. Under most conditions, the useful range of a 10 cm-wavelength (S-band) radar is considered to be close to 180 km, even though the WSR-88D system produces rainfall estimates up to 230 km. As distance increases, the beam becomes increasingly higher above average ground level because the radar's lowest elevation angle is 0.5°, the earth is curved, and there is atmospheric refraction. X-band weather radars are also used for precipitation estimates and have even shorter wavelengths (2.5 to 4 cm). The range of X-band radar is much shorter than those of C- and S-band radars. Rainfall estimates from the C-, S-, and X-band radars are processed to correct the effects of (1) ground echoes, (2) beam blockage, and (3) attenuation of the radar beam.

Radar data that provide rainfall amount products can be created at any spatial resolution. However, three primary spatial resolutions, 4 km × 4 km (2.49 mi × 2.49 mi), 2 km × 2 km (1.24 mi × 1.24 mi), and 1 km × 1 km (0.39 mi × 0.39 mi), are commonly used in the industry. The radar rainfall data are limited by the measurement of raindrop reflectivity, which is affected by factors such as raindrop size and signal reflection by other objects. Because the reflected signal measured by the radar is proportional to the sum of the sixth power of the diameter of the raindrops in a given volume covered, small changes in the size of raindrops can have a dramatic effect on the radar's estimate of rainfall (Pathak 2013). For this reason, and owing to overshooting beams and poor radar calibration, radar-derived rainfall data are scaled to match the volume measured at coincident rain gauges using bias adjustment techniques.

Radar rainfall outputs are generally in the native polar coordinate system of the radar or resampled to a grid in Cartesian coordinates. Aggregation or disaggregation of gridded radar rainfall data is often necessary with grid-based distributed models because the model grid is not at the same resolution as the radar rainfall data input or it has a different geographic projection. Besides file-format manipulation, the link between radar rainfall data and hydrologic models requires spatial aggregation from one grid to another or to subbasin areas. Aggregation from polar coordinates to basins or to rectangular grids is usually accomplished with the help of a geographic information system (GIS) or special-purpose spatial analysis tools that can handle spatial data in geographic projections.

High-quality radar rainfall records in the United States have been kept only since the mid-1990s. Moreover, some significant changes in computation algorithms have been implemented during that period. Fifteen to twenty years of radar rainfall data have limited utility for rainfall frequency analyses or long-term statistical studies. However, short-term storm analyses can be performed using radar rainfall data for a specific, recent historical event, coupled with continuous hydrologic modeling. Although radar rainfall data are increasingly available over the World Wide Web, the data processing, quality assurance/quality control, and calibration using rain-gauge data all require certain skills that are not widely available in the engineering community. At present, those skill sets are limited to a small number of firms or organizations specializing in radar rainfall estimation for hydrologic applications. Use of governmental unadjusted radar data without the proper understanding or the application of tools for quality control and correcting for bias could lead to serious errors in hydrologic analysis. The methods of estimating rainfall from radar have changed appreciably since the original deployment of the weather radar network, and these methods will continue to evolve. For example, statistical biases can change over time as algorithms are refined. This poses a major challenge in the proper interpretation of radar-derived rainfall estimates in long time series.

Radar estimates of rainfall rates depend on assumptions about the number and sizes of raindrops in a representative volume of the atmosphere. Various radar reflectivity intensity (Z) and rainfall rate (R) relationships (commonly referred to as Z–R relationships) have been derived from a theoretical or empirical basis. Depending on storm type and the power of the radar, a range of Z–R relationships is possible. Once an appropriate Z–R relationship is selected, a comparison with rain-gauge accumulations is carried out to remove any systematic error, known as bias. After bias removal, differences between radar rainfall and gauge rainfall estimates remain as random error.

Understanding real-time and post-analysis quality control and post-processing yields improve radar rainfall data for hydrologic applications. Some errors tend to average out over a watershed area, although others may dramatically increase prediction errors. Dramatic progress has been made in the past decade toward achieving representative rainfall measurement through a variety of methods. More promising methods include dual-polarization and post-processing algorithms to account for attenuation and other artifacts in short-wavelength radars such as X- and S-band radars.

Distributed hydrologic models are particularly well suited to utilizing radar rainfall. Distributed models designed from the outset to utilize high-resolution rainfall rates from multiple sensors (radar, satellite, and rain gauge) allow detailed predictions at almost any location in a watershed characterized with geospatial data relating to topography, soils, and land cover. Besides hydrologic design, radar rainfall also provides timely inputs to operational decisions for water control structures and flood warning systems and guide emergency management personnel in taking proactive steps to protect life and property from flooding.

1.3 SCOPE

This manual contains nine chapters. Chapter 1 is an introduction to radar rainfall estimation. Chapter 2 provides temporal and spatial characteristics of radar rainfall data. Chapter 3 provides details on data processing as used by the NWS in estimating radar rainfall data. Chapter 4 provides a method for data quality evaluation and improvement of radar rainfall data. Chapter 5 discusses use of radar rainfall data in hydrologic modeling. Chapter 6 provides examples of estimating radar rainfall data, data analysis, and data applications. Chapters 7, 8, and 9 discuss advanced topics on a framework for bias analysis of radar data; rain-gauge rainfall data augmentation and radar rainfall data analysis; and design of rainfall monitoring networks.

1.4 AVAILABILITY OF RADAR RAINFALL DATA WITHIN THE UNITED STATES

Climate Data Online (CDO) provides access to the National Centers for Environmental Information (NCEI, formerly NCDC)'s archive of global historical weather and climate data in hourly, daily, monthly, seasonal, and yearly time steps. Other regional or state agencies may have their own data collection system and should be checked as well.

The PRISM (parameter-elevation regressions on independent slopes model) climate group gathers climate observations from a wide range of monitoring networks and develops a spatial climate data set that contains short- and long-term climate patterns that cover from 1985 to the present. Daily, monthly, and yearly precipitation and temperature grids for the entire United States are available for download. PRISM volumes can be combined with observed hourly rainfall gauges to disaggregate daily volumes into hourly grid cells.

RFC (River Forecast Center, NWS) offices can be contacted to obtain radar rainfall data [multisensor precipitation estimator (MPE)]. Figure 1-4 shows the RFC radar extent. This map can be used to identify the geographic location(s) covered by the RFC. MPE integrates precipitation information from rain gauges, radar, and satellites to generate gridded precipitation fields in the hydrologic rainfall analysis project (HRAP) grid format with a spatial resolution of 4 km × 4 km (2.49 mi × 2.49 mi). MPE data are used at the RFCs and WFCs (Weather Forecast Centers, NWS) throughout the nation. The advantage of using hourly MPE data is the flexibility of isolating a

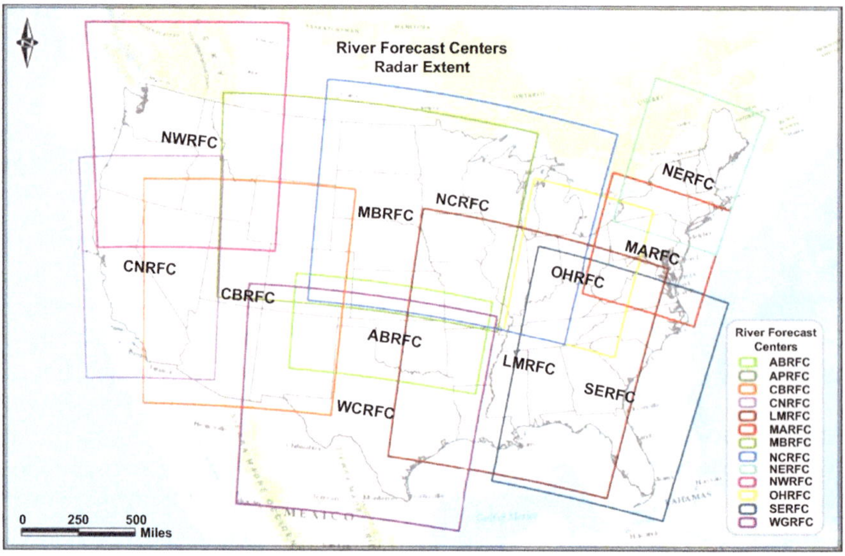

Figure 1-4. River forecast centers radar extent.

specific period to pinpoint the timeline of a storm. It should be noted that the MPE data uses Zulu Time or UTC (universal time coordinated).

The following references can be used for rainfall data collection (at the time of publication of the manual):

- NEXRAD: https://www.ncdc.noaa.gov/data-access/radar-data/nexrad;
- Webmap: https://gis.ncdc.noaa.gov/maps/ncei/radar.
- NWS: http://water.weather.gov/precip (2005 to present).
- NOAA Central Library of US Daily Weather Maps: https://library.noaa.gov/Collections/Digital-Collections/US-Daily-Weather-Maps/. This site has weather maps archived from 1871 to 2002.
- Climate Data Online: https://www.ncdc.noaa.gov/cdo-web.
- PRISM Climate Group: http://www.prism.oregonstate.edu.

REFERENCES

Bras, R. L., and I. Rodríquez-Iturbe. 1985. *Random functions and hydrology.* Boston: Addison-Wesley, 405–424.

Chow, V. T., D. R. Maidment, and L. W. Mays. 1988. *Applied hydrology.* New York: McGraw-Hill.

Pathak, C. S. 2013. "Special issue on radar rainfall data analyses and applications." *J. Hydrol. Eng.* 18(2), 131–132.

Pathak, C. S., R. Teegavarapu, D. Curtis, and C. Collier. 2017. "Special issue on radar rainfall and operational hydrology." *J. Hydrol. Eng.* 22(5), E2017001.

Seo, D. J., E. Habib, H. Andrieu, and E. Morin. 2015. "Hydrologic applications of weather radar." *J. Hydrol.* 531: 231–233.

Viessman, Jr., W., T. E. Harbaugh, and J. W. Knapp. 1972. *Introduction to hydrology.* New York: Intext Educational Publishers, 163.

CHAPTER 2

RADAR RAINFALL DATA: TEMPORAL AND SPATIAL CHARACTERISTICS

2.1 NATIVE RADAR DATA RESOLUTION

The native resolution of radar rainfall estimates derived from WSR-88D radars is 0.25 km range/0.25° azimuth relative to the radar transmitter location (Crum et al. 2010) from the legacy resolution of 1 km range/1° azimuth (NOAA 2009). Figure 2-1 shows a schematic of the legacy resolution.

Each 1 km/1° (or 0.25 km/0.25°) element is defined as a pixel. The rainfall estimate at each pixel is represented as the average rainfall estimate over the pixel area. Because the native radar resolution is expressed in polar coordinates, the pixel size increases with increasing distance from the source radar. For example, at a 20 km (12.4 mi) range, the pixel size is about 8.8 ha (21.8 ac). At 120 km (74.6 mi), the pixel size increases to about 210 ha (519 ac). Depending on the specific radar scan strategy in operation, NWS Doppler 88D radars scan at approximately 4.5-, 5-, 6-, or 10-min intervals. For most rainfall events of hydrologic interest, scan rate intervals are typically 4.5, 5, or 6 min (NOAA 2009).

The high resolution provided by radar rainfall data offers many more observations from which a rainfall surface is defined. For example, instead of one rain-gauge observation per 25 km^2 (~10 mi^2), 0.25 km/0.25° resolution radar data offer nearly 100 rainfall observations in the same area at a distance 60 km (37 mi) from the radar source. Fine-scale spatial resolution improves the representation of the true rainfall surface, which in turn enables more accurate estimates of watershed volumetric inflows at each time step.

Raw radar rainfall estimates in the form of reflectivity values from individual NWS Doppler 88D radars are freely available from the NCEI radar archives (https://www.ncdc.noaa.gov/data-access/radar-data). The NCDC

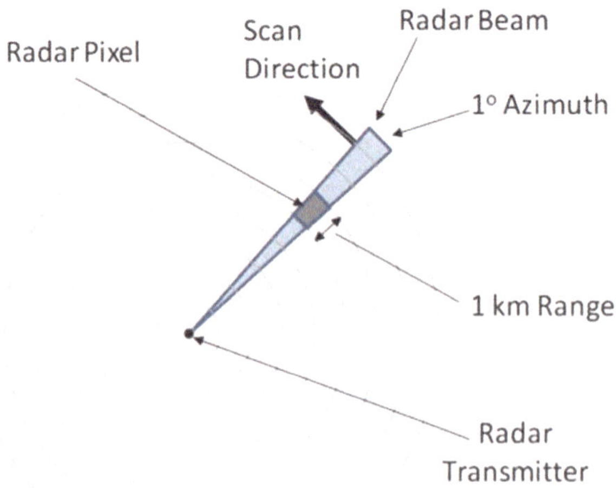

Figure 2-1. Radar pixel schematic.

Weather and Climate Toolkit is also available for downloading at https://www.ncdc.noaa.gov/wct/. The Toolkit allows users to visualize and export data to common scientific formats such as Shapefile, Arc/Info ASCII Grid, Gridded NetCDF, and more.

2.2 RADAR RAINFALL DATA MOSAICS

Radar rainfall estimation is subject to several issues associated with individual radars. For example, the radar beam elevation increases with increasing distance from the source radar. As distance increases beyond about 100 km (62 mi) from the source radar, the radar data is less and less useful. Because of the curvature of the earth, the radar beam scans even higher segments of the atmosphere, often overshooting the rain and causing underestimation. Heavy rainfall at or near the radar transmitter attenuates the radar signal, which can also cause underestimation. Obstructions such as buildings, trees, and mountains can block the radar signal from reaching its target rainfall.

To overcome these problems and others, data from nearby radars with overlapping coverage (see Figure 2-2) are combined to form a single-radar data mosaic. Radar mosaics provide continuous images of storm events over regional and national scales. Because of the widely varying pixel sizes associated with polar coordinates and the cumbersome nature of working with polar coordinate data sets, where each radar image is centered on a differ-

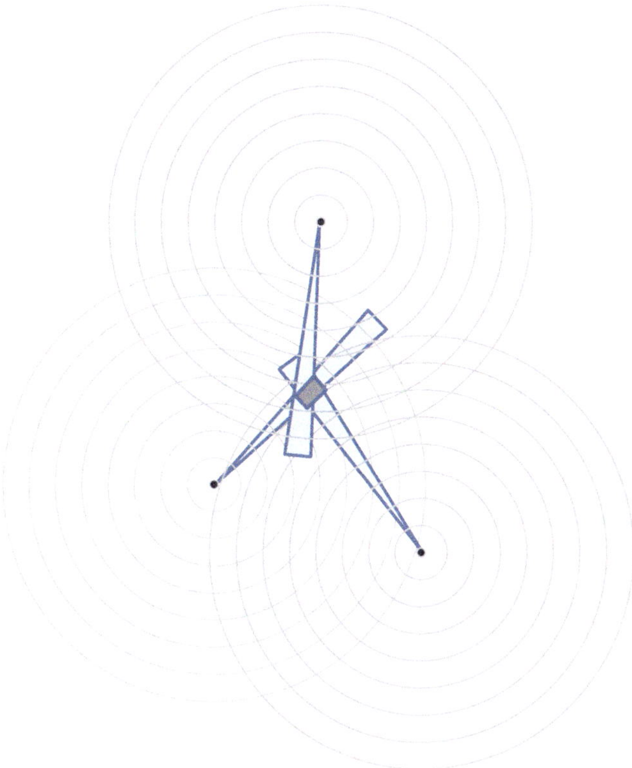

Figure 2-2. Overlapping radar coverage.

ent radar site; polar data are typically remapped to a rectilinear grid system. Figure 2-3 shows an example of NWS-enhanced radar images combining data from all available Doppler 88D radars in the continental United States remapped to a rectilinear grid.

2.3 DATA FORMATS AND RESOLUTIONS

Common scientific formats for gridded radar rainfall data include ESRI Shapefile, Arc/Info ASCII Grid, Gridded NetCDF, and others. The radar rainfall estimates are typically manipulated using tools found in a GIS. The most common gridded data resolutions include

- 5 min, 1 km × 1 km (0.39 mi × 0.39 mi),
- 15 min, 2 km × 2 km (1.24 mi × 1.24 mi), and
- 1 h, 4 km × 4 km (2.49 mi × 2.49 mi).

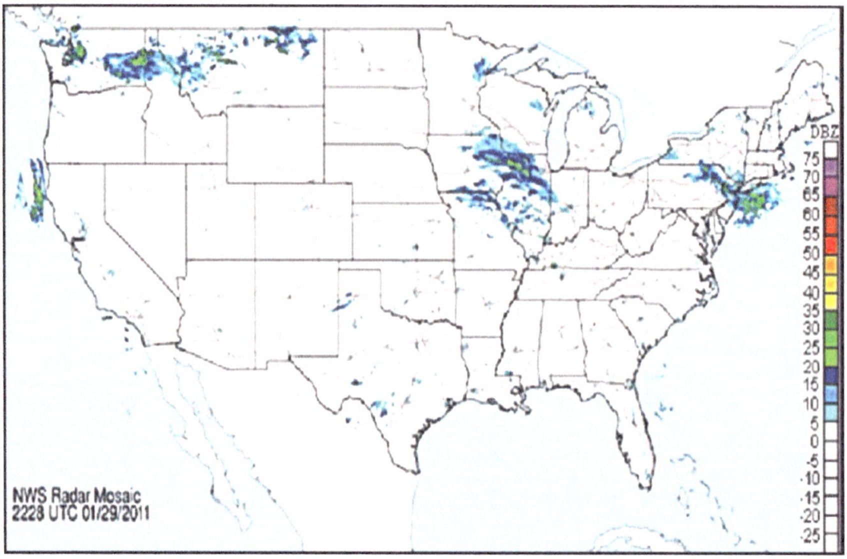

Figure 2-3. National Weather Service enhanced radar image.

However, higher resolutions (i.e., 1 min, 0.25 km) will soon be commonly available. Commercially produced gridded rainfall data sets are available in a variety of temporal and spatial resolutions. The data are typically available in GIS-compatible formats, which are easily processed for input to hydrologic models. Of special note are two standardized grid systems in common use by US government agencies: the Hydrologic Rainfall Analysis Project (HRAP) grid, used by the US NWS; and the standard hydrologic grid (SHG), used by the US Army Corps of Engineers (USACE).

2.3.1 Hydrologic Rainfall Analysis Project Grid

The HRAP grid is a square-celled map grid based on a polar stereographic projection using a spherical earth datum [HEC (Hydrologic Engineering Center) 2006]. The HRAP grid consists of 4.7625 km cells with the y-axis aligned with the central meridian (105°W) and the x-axis aligned with the standard parallel (60° N) (Figure 2-4). The following parameters describe the grid:

- Units: .... meters,
- Datum:......................... sphere (radius = 6371.2 km),
- Standard parallel: ...... 60° 0′ 0″ North, and
- Central meridian:....... 105° 0′ 0″ West.

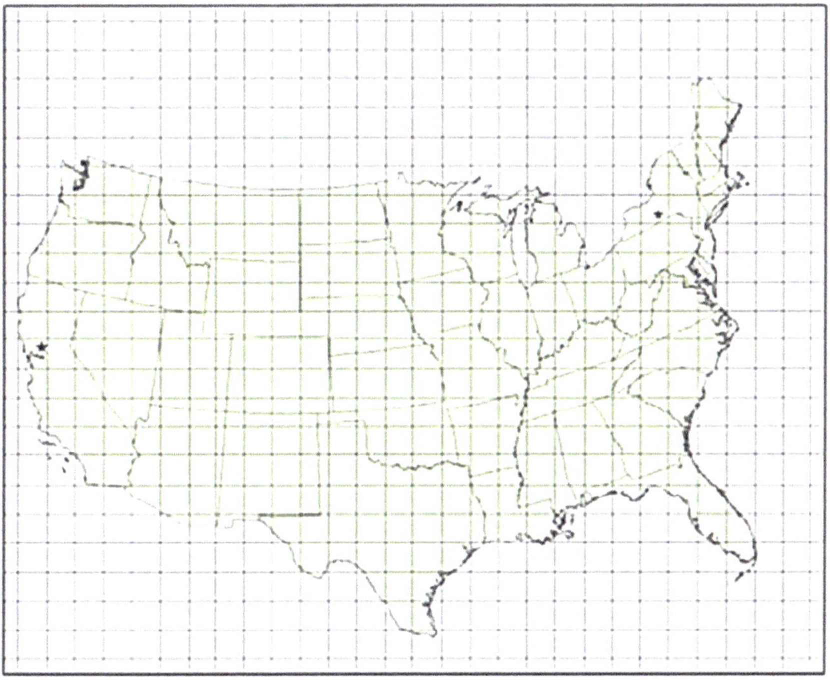

Figure 2-4. HRAP Grid projected over the continental United States.

HRAP gridded data is difficult to use in conjunction with other geospatial products because most geospatial data are in a geodetic datum coordinate system such as WGS84 and NAD83 (based on an ellipsoidal earth). The NWS website (https://www.ncdc.noaa.gov/wct/) provides instruction on how to use NEXRAD data in ArcView or ArcInfo GIS formats.

2.3.2 Standard Hydrologic Grid

The SHG is based on the Albers equal-area conic map projection developed by USACE HEC (HEC 2006). Figure 2-5 shows the SHG projected over the continental United States. The SHG is available in multiple resolutions to allow the user to select a resolution suitable for the hydrologic study needed. These resolutions are 10,000 m, 5,000 m, 1,000 m, 500 m, 200 m, 100 m, 50 m, 20 m, and 10 m. The following parameters describe the grid:

- Units: ....meters,
- Datum:...............................North American Datum, 1983 (NAD83),

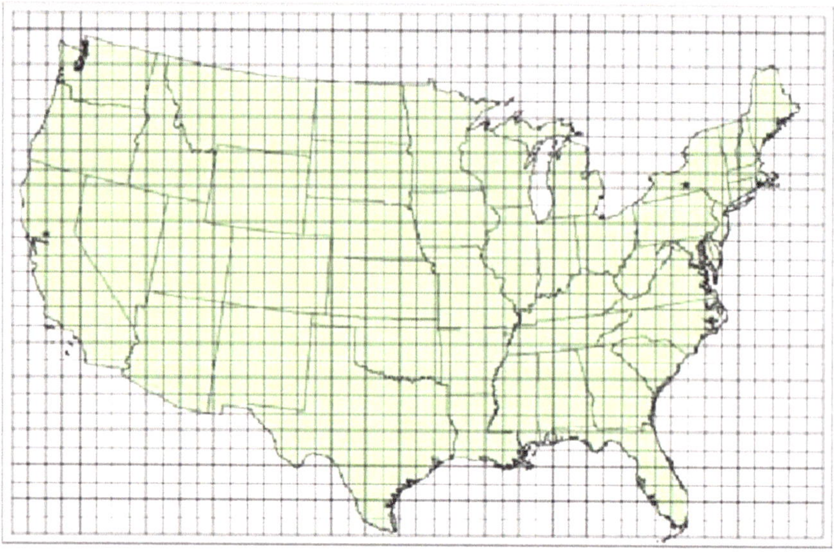

Figure 2-5. SHG projected over the continental United States.

- 1st Standard parallel: ....29° 30′ 0″ North,
- 2nd Standard parallel:......45° 30′ 0″ North,
- Central meridian:..............96° 0′ 0″ West,
- Latitude of origin:.............23° 0′ 0″ West,
- False Easting:......................0.0, and
- False Northing:..................0.0.

The SHG was proposed as an alternative to HRAP for use in hydrologic models where other GIS data would be used. The Albers-based coordinate system was selected because it matched several nationwide data sets developed by USGS (US Geological Survey) and other federal agencies. Utility programs developed by HEC to load NEXRAD data into HEC-DSS files for use in HEC's hydrologic modeling programs store grids either in their original HRAP geometry or in SHG geometry. The conversion to SHG is accomplished using area-weighted averaging.

2.4 RADAR RAINFALL DATA QA/QC AND DATA MANAGEMENT

There are several potential issues with radar data of which users need to be aware. Although there are automated methods for checking and verifying radar data from either commercial or government sources, it is always good practice to visually inspect the data. As a general rule, if some-

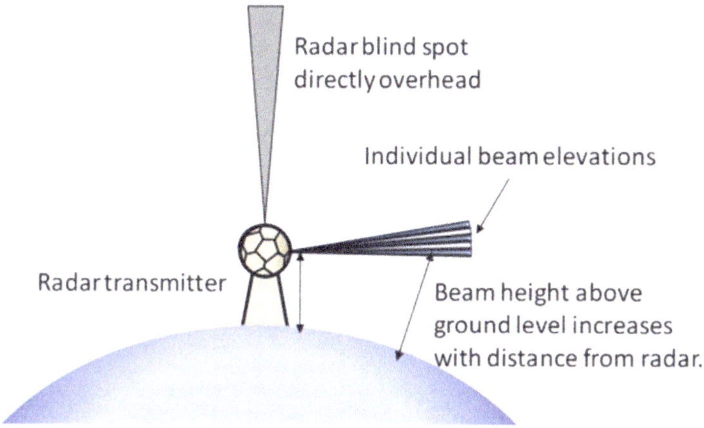

Figure 2-6. Radar beam considerations.

thing does not look natural, it probably is not. True rainfall patterns are complex and variable. If the apparent rainfall variation is too sharp—to the point of discontinuity—something "unnatural" is likely the cause and should be investigated. If a feature is too linear or geometric, again, it should be investigated and corrected if possible. It is important to investigate the quality of the radar data over the target watershed. Frequently, many anomalies may appear in a set of radar images, but if they never appear over the target watershed, the data could be used with confidence. Here are a few examples of data anomalies to watch for (see Figures 2-6 to 2-10 for visual representations):

- Beam blockage—The radar beam is fully or partially blocked by a solid object in the path of the beam (e.g., building, tower, or mountain). Often seen as a narrow wedge-shaped feature in the radar image.
- Ground clutter—The radar beam is reflected by nearby buildings or hills. This is often seen as "speckles" of apparent high-intensity radar intensities near the radar transmitter.
- Below-beam effects—These include strong horizontal winds below the radar beam, evaporation below the radar beam, and coalescence below the radar beam (see Figure 2-7).
- Elevated radar sites—Many WSR-88D Radar Data Acquisition Unit sites are located on higher terrain to reduce the impact of beam blockages. These sites are more prone to below-beam effects than sites that are located at lower locations relative to their surroundings. This difference is important to remember when comparing rainfall effects from adjacent radars whose site elevations may be thousands of feet different.

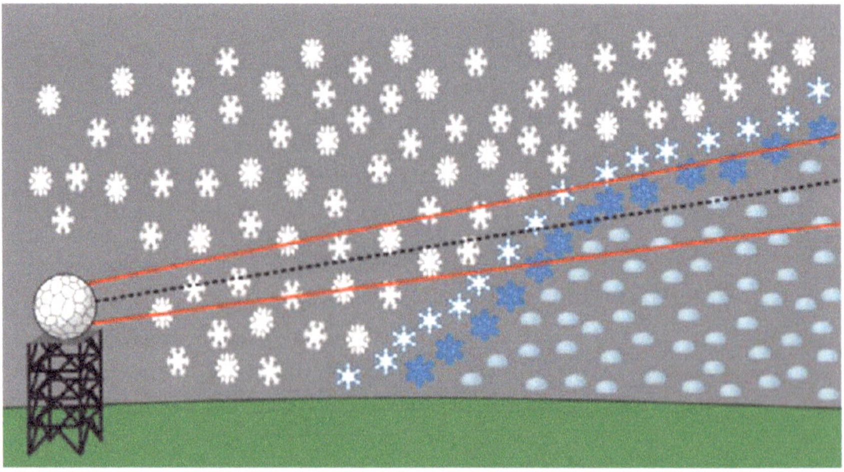

Figure 2-7. Example of below beam effects (coalescense).

- Radar coverage limit discontinuity—Overlapping radars often have slightly different characteristic signatures because of equipment and electronic calibration differences. The differences may be small, but when integrated over time, they appear as a discontinuity at the limit of radar coverage.
- Hybrid scan discontinuity—The lowest possible beam elevation is preferred for rainfall estimation. However, close to the radar transmitter, the lowest beam may encounter too much ground clutter to be valuable. In this case, the next higher available beam is substituted. A circular discontinuity is often visible at the point where the transition from one elevation to the next occurs.
- Overhead "blind" spot—The NWS Doppler 88D radars cannot see directly overhead, creating a blind spot or "cone of silence" over the radar site.
- Bright bands—Bright bands are caused when the radar beam intersects the freezing layer in the atmosphere. Wet frozen precipitation has a significantly higher reflectivity than either rain or snow. When snow in the atmosphere falls through the freezing layer, snow turns to a wet slushy mix, which generates very strong radar echoes or "bright bands" on the radar image, suggesting high-intensity rainfall. The bright band often appears as a bright circle or arc centered on the radar site. The diameter of the bright band or arc depends on the elevation of the freezing layer intercepted by the radar beam. Hail is also "wet ice" and produces very intense radar echoes and suggests anomalously high rainfall rates, although the signal is not always as distinctive as bright bands.

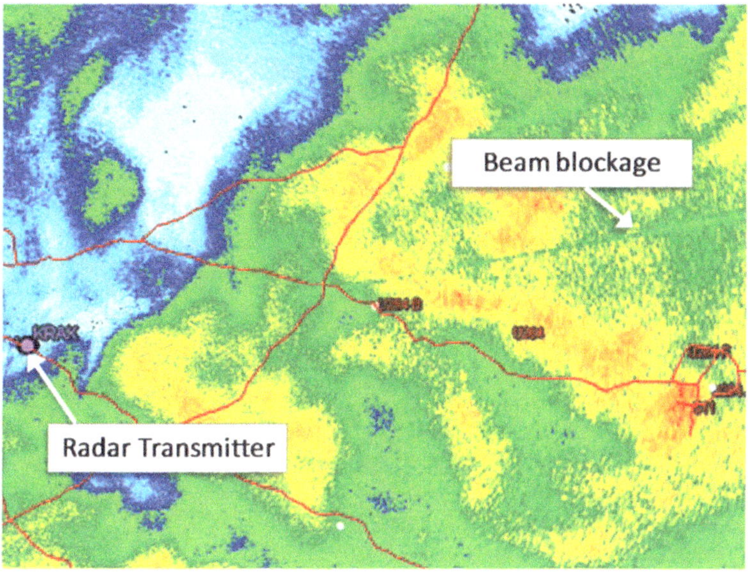

Figure 2-8. Example of radar beam blockage.

- Range fade—As the distance from the source radar beam increases, the beam elevation above ground increases, and the beam overshoots the rainfall, causing underestimation.

Figure 2-8 shows an example of beam blockage exhibited by a long, thin wedge discontinuity in the radar image. Extending the "wedge" axis to the west intersects exactly with the radar source. The likely cause is a tall, slender structure such as a communications tower or a smokestack along the path of the radar beam. Some anomalies such as the example of radar beam blockage, shown in Figure 2-8, are readily apparent in individual images at each time step. Others are more subtle and only become noticeable after a long-term integration on the order of days to months. Figure 2-9 shows a mosaic of multiple radars over the Florida peninsula for 12 months. Several types of anomalies are present, including radar limit discontinuities, hybrid scan discontinuities, ground clutter, and range fade.

Circular radar limit discontinuities appear at the 230 km range limit of each radar coverage. Hybrid scan discontinuities are circular features with diameters of about 30 km centered on the radar locations, as shown in Figure 2-9. Ground clutter, shown in Figure 2-9, appears as isolated spots of very high intensity, significantly different from adjacent pixels. Range fade is seen over the Gulf of Mexico as the radar colors transition from dark blue to white, not because there was little or no rain but because at large distances, the radar beam overshoots the rainfall.

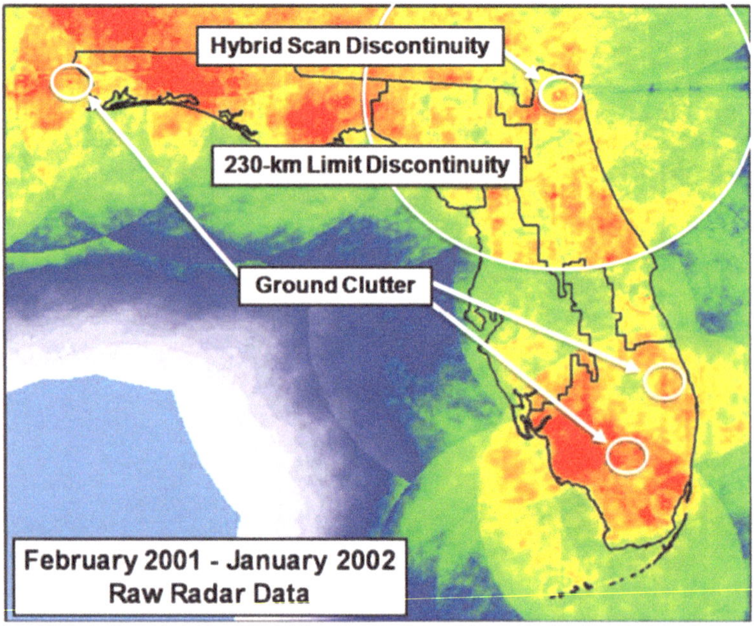

Figure 2-9. Examples of different radar image anomalies.

The discontinuities presented in Figures 2-8 and 2-9 should be corrected before use, especially if the anomalies impact target watersheds. An example of radar coverage limit or extent of discontinuity before and after correction is shown in Figure 2-10. The extent of discontinuity can be mitigated using GIS tools through an iterative adjustment, up or down, of portions of the radar data field until it creates a seamless image. Eliminating small areas of ground clutter speckles is accomplished by deleting the anomalous pixels and performing simple interpolations using adjacent pixels.

2.5 GAUGE-ADJUSTED RADAR RAINFALL ESTIMATES

After correcting radar anomalies to the greatest extent possible, the raw radar data should be compared with local rain-gauge observations to determine if the radar data field is biased relative to the gauge data. If the radar data are biased, the radar data set should be corrected or calibrated to produce gauge-adjusted radar rainfall (GARR) estimates that are consistent with rain-gauge observations. Radar rainfall calibration procedures are discussed in later chapters.

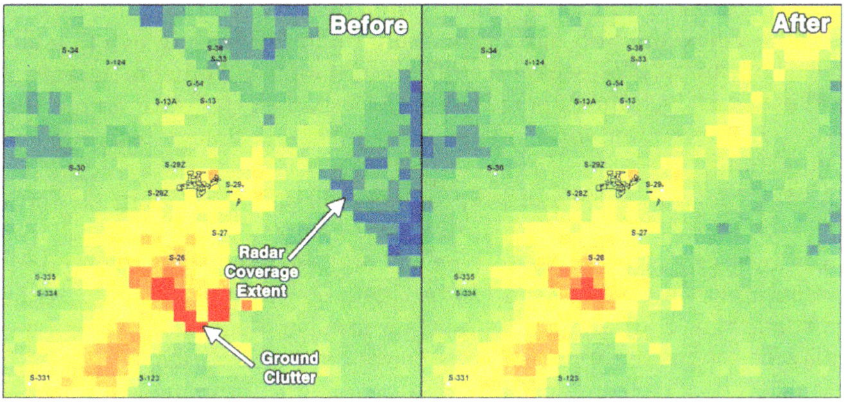

Figure 2-10. Example of radar coverage limit or extent of discontinuity before and after correction.

Note that while rain-gauge observations are often considered "ground truth" when compared to radar rainfall estimates, radar data are equally valuable in checking the validity of rain-gauge observations. For example, a partially plugged rain gauge may yield a valid storm total but an incorrect time distribution (a partially plugged gauge gradually drains over time, eventually measuring the correct total). Comparing the accumulation curves derived from the rain-gauge reports with the radar-estimated rainfall accumulation curves at the gauge location often identifies the problem. If a gauge is plugged, the gauge is eliminated from the radar calibration process. However, if the gauge total looks reasonable, the gauge total can be distributed using a normalized radar rainfall accumulation curve.

It is always desirable to have rain-gauge observations within the target watershed area. However, that is not a necessary condition to calibrate radar rainfall estimates. In reality, radar rainfall calibration is not calibrating rainfall per se. It is calibrating the performance of the radar. Aside from the anomalies discussed earlier, the radar "performance field" is typically much smoother than the actual rainfall field. This feature allows hydrologists the flexibility to use gauges outside the watershed of interest to calibrate radar data.

2.6 TOOL FOR RADAR RAINFALL DATA ANALYSIS (HEC-METVUE)

The USACE HEC (http://www.hec.usace.army.mil/) developed software known as HEC-MetVue (Meteorological Visualization Utility Engine) to perform radar rainfall data analyses. The software allows for the display, verification, and editing of spatial data through interactive and visual means. It processes and edits precipitation and other meteorological data sets, including operations such as spatial extents trimming, spatial translation, temporal shifting, image scaling, image resizing, and animations.

HEC-MetVue is a Java-based program built on the NetBeans Platform. The software's interface consists of a data explorer and legend panel, map windows, properties and output panels, and a variety of tools. It supports visual and mathematical comparisons between two images. It also accommodates time zone and unit settings for input and output, as well as display. The software supports the visualization, manipulation, and analysis of a variety of meteorological data types, in particular, precipitation, temperature, and snow.

HEC-MetVue supports several input data formats, including HEC-DSS, ASCII_GRID, NEXRAD, XMRG, NetCDF, GRIB, and PRISM's BIL. By reading multiple images, the software aggregates those images temporally and spatially and computes basin averages for any desired shapefiles loaded in the program. HEC-MetVue operates on triangulated irregular network (TIN) data sets constructed from either time series of gauge data inputs, time series of gridded data inputs, or TIN inputs. For output, it provides for saving multiple images, aggregate images, and re-projected images as gridded data sets (in HEC-DSS or ASCII_GRID formats), as well as TINs (the native HEC-MetVue TIN format). The software output options also include basin-average time series for select shapefiles and corresponding summary statistics logs.

The primary functions of HEC-MetVue are extreme storm analyses, development of design storms, including standard frequency events and probable maximum storms, and their preparation as input for rainfall runoff models (such HEC-HMS: hydrologic modeling system). HEC-MetVue can also be used for the calibration of radar-observed precipitation data sets and merging them with forecast precipitation as inputs for hydrologic models.

2.7 USE OF RADAR RAINFALL DATA

Radar rainfall data are used for a variety of purposes. Some of these uses include real-time hydrologic monitoring, flood forecasting, flood hazard and disaster management, probable maximum precipitation estimation, missing data estimation, extreme storm analysis, and many others. Chapter 6 provides details on the radar rainfall data uses.

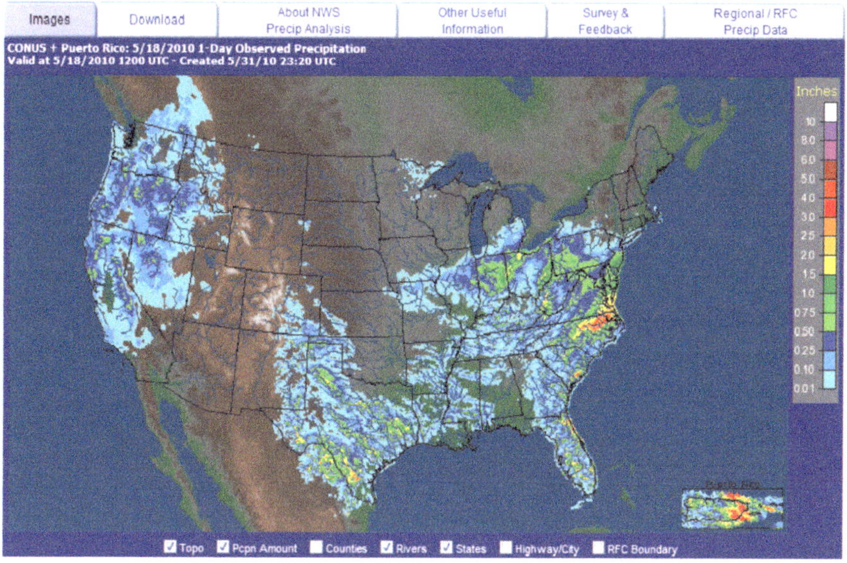

Figure 2-11. NWS precipitation analysis.

2.7.1 Real-Time Hydrologic Monitoring, Flood Forecasting, and Disaster Management

The NWS routinely uses gridded rainfall data for real-time hydrologic forecasting. Figure 2-11 shows one-day rainfall totals for May 18, 2010. Technically, the NWS product is an MPE. Nominally, it is gauge-adjusted radar rainfall occasionally enhanced with satellite rainfall estimates to overcome radar coverage gaps in the mountainous western United States and/or missing radar data (http://water.weather.gov/precip/). The NWS MPE utilizes the HRAP grid with hourly, 4 km × 4 km (2.49 mi × 2.49 mi) resolution. NWS MPE data are used in river forecast operations in watersheds east of the Rocky Mountains.

Archived NWS MPE data are often available from individual RFCs (http://water.weather.gov/ahps/rfc/rfc.php), shown in Figure 2-12, or the NCEI (https://www.ncdc.noaa.gov/data-access/radar-data).

The USACE also use NWS MPE data to forecast reservoir inflows to support flood control operations. Increasingly, regional and local agencies are using real-time gauge-adjusted radar rainfall data to support a variety of activities including flood forecasting, flood control, and stormwater management. The scale of hydrologic issues addressed at the regional and local levels requires higher-resolution gridded rainfall data than the hourly, 4 km × 4 km (2.49 mi × 2.49 mi) resolution MPE data. Higher-resolution gauge-adjusted radar rainfall data are obtained from commercial service providers.

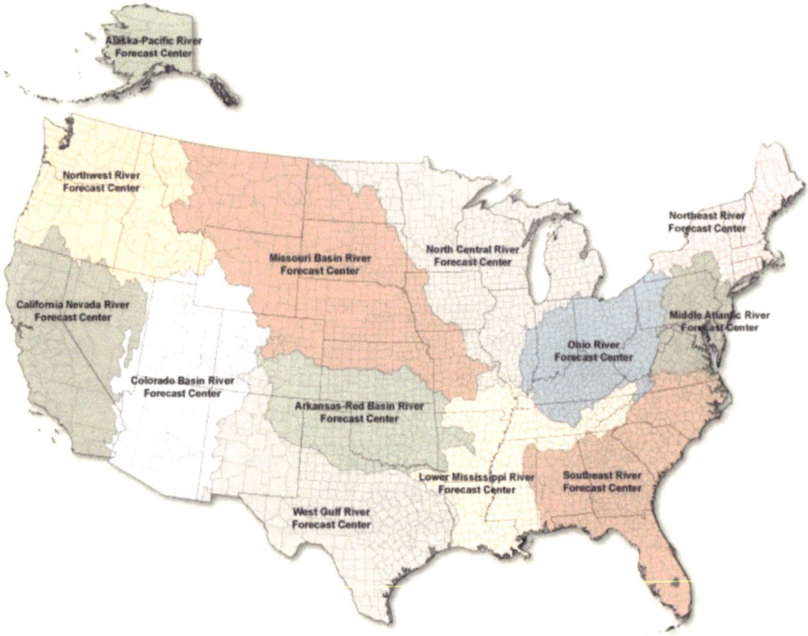

Figure 2-12. Map of NWS River Forecast Center areas of responsibility.

For example, the Louisville Jefferson County Metropolitan Sewer District in Louisville, Kentucky, uses real-time 5 min, 1 km × 1 km (0.39 mi × 0.39 mi) gauge-adjusted radar rainfall data to manage stormwater in a series of underground pipes, tunnels, and storage areas. The Rice University and Texas Medical Center in Houston, Texas, uses high-resolution gauge-adjusted radar rainfall estimates to forecast flood threats from Brays Bayou. The water management operations of the South Florida Water Management District in West Palm Beach, Florida, are supported by real-time 15 min, 2 km × 2 km (1.24 mi × 1.24 mi) resolution gauge-adjusted radar rainfall data (see Figure 2-13).

2.7.2 Probable Maximum Precipitation Estimation

Hydrological design will benefit from the use of radar data for the estimation of probable maximum precipitation (PMP). However, such estimation is often dismissed because of the shorter length of data compared to the long-term data that is available from rain gauges (Collier and Hardraker 2007). The utility of the radar data is also being questioned for use in the frequency analysis of extremely rare events because of the lack of a mini-

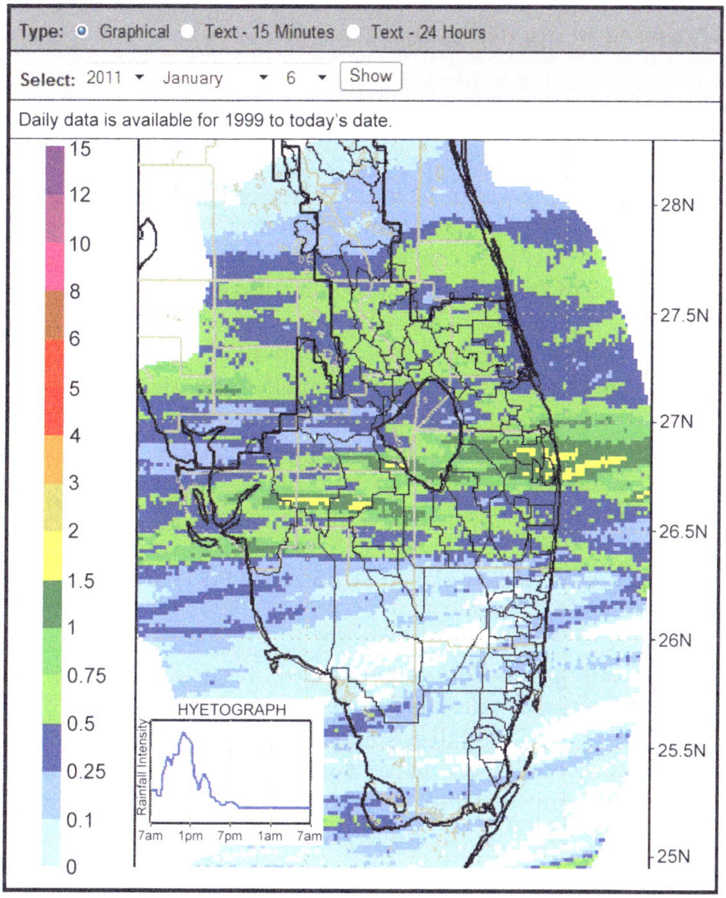

Figure 2-13. Sample daily rainfall estimates for the South Florida Water Management District.

mally reasonable data length (10 years). In spite of all these limitations associated with the available length of radar data, the data is now used extensively for extreme rainfall analysis. Collier and Hardraker (1996) developed methods to estimate maximum rainfall totals to obtain realistic estimates of PMP. Depth–area–duration curves were developed by Cluckie et al. (1987) using radar-based rainfall data. Cluckie and Pessoa (1990) characterized actual storms for the development of PMP estimates using radar-based rainfall data. Reliable radar data is now available in many parts of the world at a temporal resolution that is adequate for frequency analysis for hydrologic design. One concern with radar is that the evolving technology suggests no consistency in methods used for obtaining reflectivity data.

Often the fine temporal resolution radar-based rainfall data are more useful in analyzing short-duration storms.

2.7.3 Missing Data Estimation

The rainfall missing data from rain gauges are common occurrences and are due to rain-gauge equipment malfunction as well as malfunction of the data communication (or data transmission) system. In a given year, typically missing data vary between 5% and 10%. For example, at South Florida Water Management District, during 2005, approximately 4.2% daily rainfall data were missing due to malfunction of equipment that included both sensors and communication systems (Pathak 2008). Often, rain-gauge missing data are estimated using various mathematical methods. However, radar rainfall estimates are also used in filling the rain-gauge missing data with the superior data quality. Details on missing data estimation are presented in Chapter 8.

2.8 RADAR RAINFALL DATA ISSUES AND FUTURE PERSPECTIVES

The value of high-resolution spatiotemporal estimates matched with equally high-resolution hydrologic models is apparent. The pace of the introduction of radar-based rainfall estimation is expected to accelerate in coming years. New technologies such as dual polarization and a phased array radar system will improve the quality of raw radar rainfall estimates. Perhaps most important, the growing database of radar rainfall estimates nationwide will open entirely new lines of research to investigate and understand the geospatial properties of rainfall events.

2.8.1 Dual Polarization

WSR-88D sites across the nation have upgraded to polarimetric radar, which adds vertical polarization to the horizontal radar waves, to more accurately discern what is reflecting the signal. Dual polarization allows the radar to better distinguish between rain, hail, and snow, which horizontally polarized radars cannot accurately do. Early trials have shown that rain, ice pellets, snow, hail, birds, insects, and ground clutter all have different signatures with dual polarization, which could mark a significant improvement in forecasting winter storms and severe thunderstorms. The deployment of the dual-polarization capability (Build 12) to NEXRAD sites began in 2010 and was completed in early 2013. The Vance Air Force Base (AFB) radar was the first operational WSR-88D to be modified to dual polarization (operational March 3, 2011).

2.8.2 Phased Array

Beyond dual polarization, the advent of phased array radar will probably be the next major improvement in severe weather detection. Its ability to rapidly scan large areas would give an enormous advantage to radar meteorologists. However, any large-scale installation by the NWS is unlikely to occur before 2020. The NOAA Severe Storms Laboratory predicts that a phased array system will eventually replace the current network of WSR-88D radar transmitters (http://www.nssl.noaa.gov/tools/radar/mpar/).

2.8.3 Archived Radar Rainfall Data

Archived radar data present a rich opportunity for new research and engineering practice. For much of the United States, historical radar rainfall estimates are available with records approaching 20 years. Although 20 years is insufficient for many statistical analyses based on annual maxima, for example, ample data exist to evaluate the geospatial properties of storms. Depth area adjustment factors (DARFs) that came into common usage were published more than 50 years ago (US Weather Bureau 1957). Historical archives of radar rainfall data are available that can support a completely new redefinition of DARFs, which will bring a much-needed refinement to engineering design (Curtis et al. 2011).

To date, the mining of radar rainfall data archives for new insights into the spatial structure of storms and how this structure impacts hydrologic design is in its infancy. It is entirely possible that insights mined from large radar rainfall data sets could ultimately impact every hydrologic design standard in the United States.

2.9 CONCLUSIONS

Accurate estimation of volumetric watershed outflow is a key objective in hydrologic analysis. Increasingly, rain-gauge observations are merged with radar rainfall estimates to produce GARR. The increased spatial resolution of the rainfall field provided by GARR can improve the estimates of watershed inflows which, in turn, improve estimates of volumetric outflows. This chapter traces radar rainfall data from its polar coordinate origins at individual radars to a mosaic of data from multiple radars remapped to rectilinear grids for use in hydrologic models. Radar data like rain-gauge observations are not without errors. Several of the common radar data anomalies were described to give potential users guidance on what to look for while visually inspecting the data before use.

The NWS and the USACE are two US agencies that routinely use real-time GARR in flood forecast and flood management operations. Regional and local agencies such as the Louisville Jefferson County Metropolitan Sewer District in Louisville, Kentucky; the Rice University and Texas Medical Center in Houston, Texas; and South Florida Water Management District in West Palm Beach, Florida; are using high-resolution GARR data in stormwater management, flood warning, and water resources management in real time. Engineers are routinely using historical GARR to support hydrologic modeling efforts. New methodologies using GARR data to develop DARFs—a parameter commonly used in hydrologic design—are under development. The growing national database of radar rainfall estimates represents the first opportunity for hydrologists to conduct in-depth studies of the geometric properties of rainfall events. Millions of individual storms cells are represented in the radar archives. Much work needs to be done, but the early results are promising and show the potential for historical GARR estimates to impact nearly all hydrologic design standards in the United States.

REFERENCES

Cluckie, I. D., P. P. Ede, M. D. Owens, and A. C. Bailey. 1987. "Some hydrological aspects of weather radar research in the United Kingdom." *Hydrol. Sci. J.* 32(3), 329–346.

Cluckie, I. D., and M. L. Pessoa. 1990. "Dam safety: An evaluation of some procedures for design flood estimation." *Hydrol. Sci. J.* 35(5), 547–569.

Collier, C. G., and P. J. Hardaker. 1996. "Estimating probable maximum precipitation using a storm model approach." *J. Hydrol.* 183(3-4), 277–306.

Collier, C. G., and P. J. Hardaker. 2007. "Using radar in hydrometeorology." In *Weather radar: principles and advanced applications*, P. Meischner (ed.). New York: Springer.

Crum, T., D. Horvat, C. Horvat, C. Calvert, M. Istok, S. DelGreco, et al. 2010. "4.6 NWS WSR-88D and TDWR-SPG data collection and distribution network: Status and plans." Preprints 16th Conf. on Interactive Information Processing Systems. Boston: American Meteorological Society.

Curtis, D. C., J. Humphrey, and D. W. Bare. 2011. "Radar-based depth area reduction factors for Colorado." Presented at the American Geophysical Union Fall Meeting, San Francisco, December 5–9.

HEC (Hydrologic Engineering Center). 2006. *GageInterp User's Manual.* Washington, DC: US Army Corps of Engineers.

NOAA (National Oceanic and Atmospheric Administration). 2009. "Federal meteorological handbook No. 11: Doppler radar meteorological observations, Part A, system concepts, responsibilities, and procedures." *FCM-H11A-2009*, Silver Spring, MD.

Pathak, C. S. 2008. "Hydrologic monitoring network of South Florida Water Management District—Appendix 2-1." *South Florida environmental report*. Accessed August 13, 2018. http://my.sfwmd.gov/portal/page/portal/pg_grp_sfwmd_sfer/portlet_subtab_draft_rpt/tab19853153/appendices_2008/v1_app_2-1.pdf.

US Weather Bureau. 1957. Rainfall intensity-frequency regime." Weather Bureau Technical Paper No. 29. Washington, DC.

CHAPTER 3
RADAR RAINFALL DATA PROCESSING

3.1 BACKGROUND

3.1.1 Programmatic Background and Organization of This Chapter

In the United States, the existing archive of data from the Weather Surveillance Radar Doppler (WSR-88D) and the Terminal Doppler Weather Radar (TDWR), available near some locations since 1991 and at most places since 1996, can be used in combination with rain-gauge observations to obtain hourly precipitation estimates. Although these estimates are subject to considerable random error relative to collocated rain-gauge reports, they can be adjusted to ensure long-term reliability. Radar estimates also are reliable for areal-average rainfall, for which observations from a few scattered rain-gauge sites might provide limited accuracy, particularly in situations with large spatiotemporal precipitation variability.

Automated radar processing algorithms are subject to errors in discriminating between hydrometeors and other targets, and among different types of hydrometeors, such as raindrops, snowflakes, hailstones, and graupel. Moreover, the algorithms yield estimates of water content and associated rainfall rates with considerable error. Therefore, practice in operational hydrology has been to rely on human input to apply quality control to raw radar precipitation estimates and to regularly adjust radar estimates with any available collocated rain-gauge reports (Kitzmiller et al. 2013). Such bias adjustments appreciably improve the absolute accuracy and long-term reliability of the raw radar estimates.

This chapter summarizes principles of radar quantitative precipitation estimates (QPE) as used by the NWS. It also summarizes considerations in applying radar and gauge QPE to hydrologic modeling tasks. Section 3.2

describes the processing sequence for WSR-88D and gauge–radar precipitation estimates and the basic principles of radar data acquisition and processing. Section 3.3 summarizes the physical principles behind the reflectivity-to-rain rate (*Z–R*) relationships and the limitations of this approach. Section 3.4 features a description of QPE products from the WSR-88D radar product generator subsystem. Section 3.5 briefly describes the statistical characteristics of radar QPE errors. Sections 3.6 and 3.7 contain high-level descriptions of the basic approaches to the statistical blending of radar and rain-gauge estimates. Section 3.8 discusses considerations of the relative value of radar and gauge QPE in situations in which gauge data is readily available or where radar coverage is compromised. Section 3.9 summarizes possibilities for the use of daily rain-gauge input for adjustment of subdaily estimates. Section 3.10 lists current sources for WSR-88D, gauge–radar multisensors, and rain-gauge data. Concluding remarks appear in Section 3.11.

3.1.2 Limitations on the Content of This Chapter

The reader must note that our discussion is limited almost entirely to single-polarization radar estimation. Although the WSR-88D network has recently been upgraded to dual-polarization capability, the current 20-year archive of radar data and products is based largely on single-polarization technology and theory. We anticipate that future updates to this manual will feature advice on practices in using dual-polarization information.

The discussion below is also limited to radar-based quantitative estimation of rainfall. Although radar detects all hydrometeors, there has to date been limited success in estimating the liquid content of precipitation other than raindrops, whose microwave backscatter properties are well known. The gauge–radar techniques outlined in this and other chapters do provide some scope for correctly estimating the liquid content of radar-detected snow, subject to limitations on the accuracy of gauge observations of frozen hydrometeors.

Finally, our discussion is largely limited to the WSR-88D system and certain NWS systems that were operational as of 2013. There are other radar processing packages whose output is available to the research and applications community, such as the Warning Decision Support System (WDSS, Lakshmanan et al. 2007) and the National Mosaic and Multi-Sensor QPE (NMQ) system (Zhang et al. 2011a). The NMQ (now referred to as Q3) was implemented within the NWS in 2014, as part of the Multi-Radar Multi-Sensor (MRMS) radar processing system (Zhang et al. 2016). Moreover, a multiyear reanalysis effort for historical data has been completed (Howard and Cano, personal communication, 2013; Zhang, personal communication, 2017).

3.2 DATA ACQUISITION AND PROCESSING

3.2.1 Stages in Preparation of Radar and Gauge–Radar Products

Radar precipitation products pass through multiple platforms for automated and human interactive processing (Kitzmiller et al. 2013). Because radar reflectivity, radar-only precipitation estimates, and gauge–radar merged estimates are all available to end users, it is instructive to consider the processing sequence and levels of refinement that characterize each of these stages, which are shown schematically in Figure 3-1. The first two stages are considered the basic processing of single-radar data; the third stage features manual quality control of radar products and merging of data from multiple radars; the fourth stage is a national-scale merging of regional multi-radar products.

The functional elements of a single WSR-88D unit are shown schematically in the first two stages of Figure 3-1. The radar data acquisition (RDA) subsystem consists of the radar signal generator, transmitter, receiver, and signal processor, typically housed within a building surmounted by the radar tower; the RDA also includes a computer for final signal processing and basic quality control of the basic radar moments, namely, reflectivity, velocity, and spectrum width. These moments are transmitted to a second computer, the radar product generator (RPG), which generates images and interpretive products, including precipitation estimates (steps explained in Sections 3.3 and 3.4). All basic moment data, images, and products are packaged as digital message files and transmitted to NWS field offices, national centers, and the general public via multiple dissemination systems, as explained in Section 3.10.

Further precipitation processing is executed outside the individual radar systems, in the Advanced Weather Interactive Processing System (AWIPS) maintained at NWS Weather Forecast Offices and River Forecast Centers (RFCs). Here, products from multiple radars are merged with each other and with gauge reports, as explained in Sections 3.6 and 3.7. These multi-radar, multisensor products, referred to collectively as Stage II and Stage III, are used for local and regional forecasting operations and are transmitted to regional headquarters and the National Centers for Environmental Prediction, where the products are merged into national-scale Stage IV mosaics (Section 3.10).

3.2.2 Physical Principles for Single-Polarization Radar Precipitation Estimates

The mass of precipitation within a volume of the atmosphere is roughly proportional to the amount of incident radar-generated radiation that is backscattered from that volume. An estimate of this returned power is obtained by statistical processing of the backscatter from multiple pulses

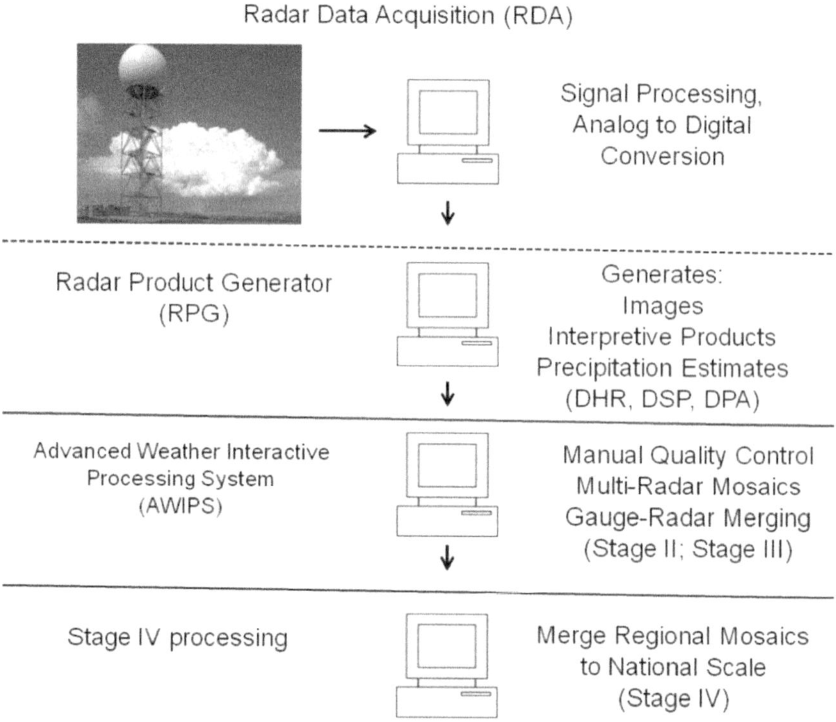

Figure 3-1. Phases in processing radar data into radar-only and radar-gauge products. The RDA and RPG phases are part of the WSR-88D system; the AWIPS and Stage IV phases are executed separately within NWS operations.
Source: Radar image (upper left), courtesy of National Oceanic and Atmospheric Administration (NOAA) Photo Library, Photo by Eric Kurth, National Weather Service.

of outgoing radio energy (Doviak and Zrnić 2006). On the assumption that the backscatters are spherical water droplets, an estimate of the water content that produces the observed backscatter is obtained as a reflectivity (Z). This value is defined by

$$Z \equiv \frac{1}{\Delta V} \sum_i D_i^6 \tag{3-1}$$

where
Z = reflectivity factor ($mm^6\ m^{-3}$),
ΔV = a unit volume, and
D_i = the diameter of the ith droplet in a unit volume (Doviak and Zrnić 2006).

Thus, the fraction of power backscattered to the radar is proportional to the sum of the 6th powers of the droplet diameters.

Of course, radar alone cannot be used to determine the spectrum of drop sizes in Equation 3-1; only the backscattered microwave power from the volume of the atmosphere is known. A variety of raindrop-size spectra, such as those documented by Marshall and Palmer (1948) can produce identical backscattering. One of the several representative drop-size distributions can be specified by applying knowledge of the meteorological conditions and the reflectivity profile in the vertical. This drop-size distribution is then applied to estimate the rain rate associated with the reflectivity through a reflectivity–rain rate (Z–R) relationship. As explained in Sections 3.2.3 and 3.2.4, the interpretation of the actual rainfall rate and liquid water content from this factor is complicated by several factors.

3.2.3 Processing for Radar Quality Control

Radar reflectivity fields feature many nonprecipitation features, and in commonly used display formats, some of these superficially resemble returns from precipitation. Backscatter from fixed ground targets such as nearby buildings, terrain, and wind farms is commonly observed and must be edited during initial data processing. Stationary ground targets and terrain can be recognized via fixed location and a zero Doppler velocity, principles that are used as a basic level of reflectivity editing (Fulton et al. 1998).

Transitory nonprecipitation features include incident solar radiation near times of sunrise and sunset (sun strobes), biological targets (Chilson et al. 2012), and external microwave sources. Among the most common nonprecipitation echoes are those from biological targets, including migrating birds, insects, and foraging bats, which are often not recognized by the current automated quality control algorithms; their returns can resemble light precipitation near radar sites [Figure 3-2(a)]. Such anomalous features are common in warm-season radar-only precipitation analyses. Although these targets can be readily detected by dual-polarization algorithms (Park et al. 2009), they are often not detected by even advanced single-polarization algorithms (Lakshmanan et al. 2010) and must be manually edited by trained analysts. The edited precipitation field for this case is shown in Figure 3-2(b).

New algorithms for quality control can be applied retrospectively to archived base reflectivity data (Zhang et al. 2011a). Examples of fields with sun strobes, migrating birds, and radar signal tests successfully removed by automated quality control are shown by Zhang et al. (2011a, Figure 2). However, it should be recognized that considerable interactive adjustment of radar and gauge precipitation fields has been carried out by RFC forecasters in preparing operational products (Figure 3-2) (Kitzmiller et al. 2013;

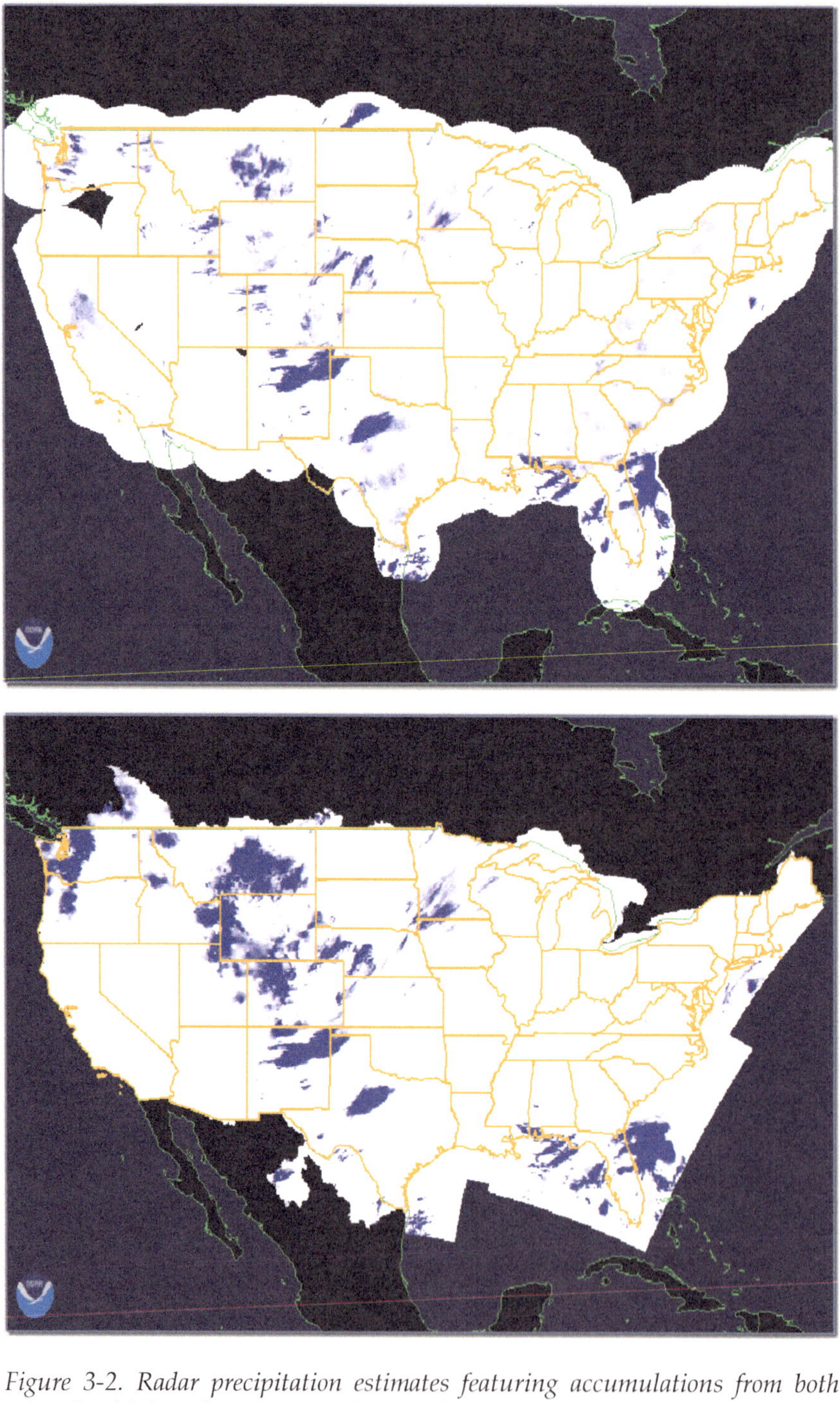

Figure 3-2. Radar precipitation estimates featuring accumulations from both migrating birds and precipitation (top) and precipitation only, the following manual quality control (bottom). Images are for 6 h accumulations ending 06 UTC, October 1, 2017.

see also Section 3.9). These precipitation products can themselves be used as a reference for quality control of contemporaneous automated products.

3.2.4 Hydrometeor Identification

Single-polarization reflectivity alone contains minimal information on precipitation phase (rain, graupel, dry or melting snow, hailstones). Because most mid-latitude precipitation initially condenses above the melting level as snow, observations of stratiform precipitation from any one radar commonly show multiple phases, typically raindrops close to the radar where the beam is well below the melting layer and snow at very long ranges, particularly in the cool season. In an intervening region, melting snow can be detected as an annular region of enhanced reflectivity (bright band) and precipitation accumulation, concentric about the radar unit (Figure 3-3). This characteristic pattern does facilitate interpretation of the reliability of the radar QPE, which is most accurate within the region closest to the radar,

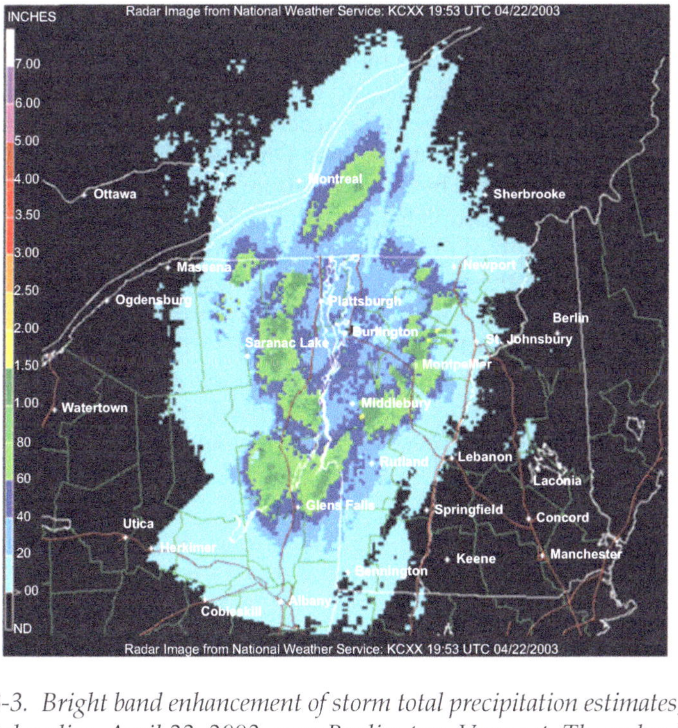

Figure 3-3. Bright band enhancement of storm total precipitation estimates, multi-hour total ending April 22, 2003, near Burlington, Vermont. The radar detected primarily rain near the radar (city of Burlington), melting snowflakes in the annular region encompassing Saranac Lake, Glens Falls, and Montpelier, and dry snow at longer ranges.

an overestimate under the bright band where melting snow is detected and an underestimate beyond the bright band where dry snow is detected.

For single-polarization radar, several algorithms to recognize and partially correct for these vertical reflectivity profile effects are available (Zhang and Qi 2010, Xu et al. 2008, Bellon et al. 2005, Seo et al. 2000); at present, none are applied in the operational WSR-88D system. Some range-dependent effects are mitigated by mosaicking of data from multiple radars, ensuring that precipitation estimates come from the radar with the lowest possible beam elevation over any one point; some are also mitigated by adjustment with rain-gauge reports.

Water-coated hailstones are highly reflective and yield a return disproportionately high for their water content. A standard procedure in single-polarization processing is to cap precipitation rates at values corresponding to a reflectivity of 51–55 dBZe on the assumption that higher rate estimates are contaminated by hail (Fulton et al. 1998).

3.3 REFLECTIVITY–PRECIPITATION RATE RELATIONSHIPS

As noted, single-polarization radar relies on assumptions about the possible range of droplet spectra that can yield a known portion of microwave radiation backscattered to the radar receiver. There are broad constraints on droplet sizes and their detectability. Droplets below 0.5 mm diameter are too small to be detected by S-band (~10 cm wavelength) radars, whereas frictional drag renders droplets larger than 5 mm too unstable to exist for any appreciable length of time without fragmenting.

However, within these limits, a wide variety of drop sizes are often observed. The reflectivity equation used to convert returned power to reflectivity factor is such that the same reflectivity can arise from a variety of liquid water contents and rainfall rates. A number of simplifying assumptions about droplet-size spectra have been used to develop reflectivity-to-rain rate (Z–R) equations. These typically take the form

$$Z = aR^b \tag{3-2}$$

where
Z = reflectivity factor in $mm^6\ m^{-3}$,
R = rainfall rate in $mm\ h^{-1}$, and
a and b = empirically derived constants.

Generally accepted values for a range from 130 to 300; those for b range from 1.2 to 1.7 (see references in the next paragraph).

As shown in Figure 3-4, some common droplet spectra characteristic of convective (thunderstorm, Smith and Krajewski 1994), cool-season stratiform (Marshall and Palmer 1948), and warm-process (tropical storm, Anag-

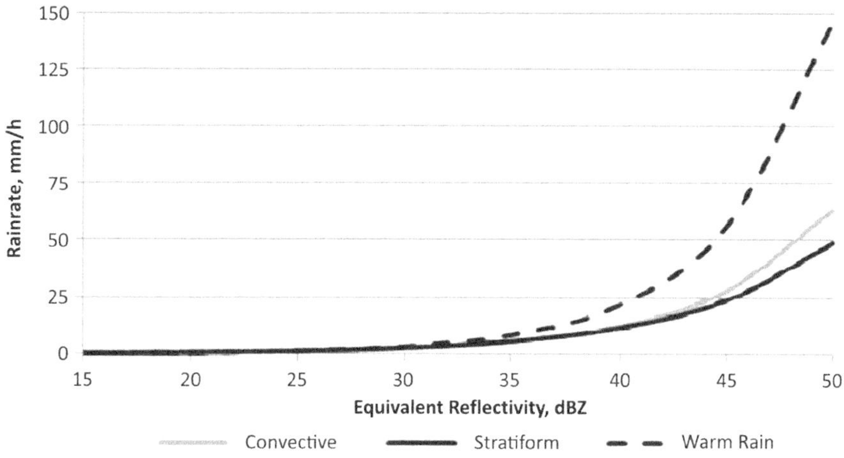

Figure 3-4. Rainfall rates associated with a range of reflectivity values for three rainfall regimes. For reflectivities over 45 dBZ, the warm rain (dashed trace) yields appreciably more rainfall than do the stratiform regime (solid black trace) and convective regime (gray trace).

nostou and Krajewski 1998) situations yield very different Z–R relationships and rainfall rates. Any one storm system often features precipitation from a combination of at least two of these three common regimes.

Within the operational WSR-88D system, Z–R relationships are adjusted manually through a control interface. Meteorologists are trained on which relationships are generally applicable in their service areas and which to choose under various circumstances. However, this approach does not account for variations in precipitation types within the radar umbrella or local temporal variations within the space of an hour or less.

It is possible to recognize the different regimes automatically through consideration of vertical reflectivity structure and knowledge of the local height of the melting layer (Xu et al. 2008). Adjustment of Z–R relationships through automated processes can yield appreciable improvement over periodic manual adjustment (Zhang et al. 2011a).

However, not all precipitation regimes fall into predefined drop-size categories, and the vertical reflectivity profile and other effects still cause substantial errors, even if the drop-size distribution at the level of the radar beam is modeled well. The most effective methods for radar precipitation estimation rely on multisensor merging, particularly through the use of collocated rain-gauge measurements. As explained in Sections 3.6 and 3.7, statistical adjustment of radar estimates to ensure better agreement with available gauge measurements can result in final estimates that improve on either radar or rain gauges alone. In current operational practice, rain-gauge information is used to the greatest extent possible.

3.4 RADAR QPE PRODUCTS FROM THE WSR-88D RADAR PRODUCT GENERATOR

The most common approach to precipitation estimation involves initial quality control to detect and remove nonprecipitation echoes, followed by vertical interpolation of data from multiple antenna elevations to yield a quasi-horizontal representation of the precipitation rate field. This procedure is followed to obtain the digital hybrid reflectivity (DHR) field, as described by Fulton et al. (1998). The DHR is generated from reflectivity input that has been automatically examined for the presence of statistical outliers, ground clutter, and anomalous propagation that have distinctive signatures distinguishable from precipitation (Fulton et al. 1998, Krajewski and Smith 2002). In turn, instantaneous precipitation rates derived from the DHR are used to create a Cartesian-gridded 1 hr accumulation product with a mesh length of ~4 km, the digital precipitation array (DPA) and the radially gridded digital storm total precipitation (DSP). These three products are used in downstream analysis operations such as gauge–radar merging. A graphical version of the DSP is shown in Figure 3-2.

3.5 ERROR DISTRIBUTION OF RADAR RAINFALL ESTIMATES

3.5.1 Errors Caused by Sampling Differences between Gauge and Radar

Particularly in intense rainfall, it is known that radar sampling limitations and differences between radar small-area and gauge point rainfall can cause major discrepancies between gauge and radar estimates. Given that the errors are rather large relative to the precipitation amounts themselves, many users wish to have an idea of the range of actual values that can be expected for any one estimate.

Gauge–radar error distributions for radar with only long-term bias correction were presented by Ciach et al. (2007), who analyzed a large multiyear data set covering Oklahoma. For DPA products, with radar-estimated accumulations ≥5 mm°h^{-1}, they determined that approximately 50% of the associated gauge reports would lie between 50% and 150% of the radar estimate (i.e., 2.5 mm and 7.5 mm for a radar estimate of 5 mm). Approximately 90% of the gauge values would fall between 10% and 200% of the radar estimate. The fractional errors become much larger for small amounts, although the associated absolute errors are themselves smaller.

Some reduction in systematic or spatially coherent error is achieved through gauge–radar merging, explained in the following sections. Aggregation of the radar estimates in time and space reduces the impact of random errors, hence the generally observed tendency of gauge and radar estimates to be better correlated with longer accumulation periods.

3.5.2 Numerical Truncation Errors in RPG Products Before 2004

In the WSR-88D record before 2004, a computational artifact in the RPG caused significant underestimation of rainfall in lighter precipitation (Fulton et al. 2003). This error had a major impact on long-term accumulations and was often not corrected by gauge–radar merging (Zhang et al. 2011b, Wang et al. 2008b). Following the implementation of an RPG software upgrade during 2003–2004, the low precipitation bias caused by the truncation error was largely corrected (Zhang et al. 2011b). Users of archived DPA products generated before 2004 should be aware of this error and are advised to consider the use of hourly and daily gauge reports in correcting it.

3.6 APPROACHES TO GAUGE–RADAR ADJUSTMENT

Most time series of radar estimates feature long-term and temporally varying biases. Gauge/radar biases ranging from 0.8 (radar overestimation) to 1.2 (radar underestimation) are commonly reported. Within the time record of any one radar umbrella, underestimation tends to occur in the cool season and overestimation in the warm season. All the approaches presented in this section are designed to correct the time varying, as well as the long-term systematic biases.

3.6.1 Mean-Field Bias Correction

Following early studies documented by Wilson (1970) and Brandes (1975), methods for adjusting radar estimates toward an agreement with rain-gauge data have been extensively studied. A very basic approach is *mean field bias correction*, which attempts to determine and correct a bias in radar electronic calibration or possibly an inaccurate Z–R relationship over the entire radar umbrella. Coincident rain-gauge reports and radar estimates are collected, and the ratio of the mean values over a recent period is determined; that is,

$$\beta_{mf} = \sum_t \sum_j \Lambda_t G_{tj} \Big/ \sum_t \sum_j \Lambda_t R_{tj} \tag{3-3}$$

where
β_{mf} = mean-field bias,
G_{tj} = gauge measurement at time–space location t-j,
R_{tj} = corresponding radar estimate, and
Λ_t = temporal weighting factor.

The time-dependent weighting factor Λ_t decreases with increasing time lag. For real-time hydrologic applications, the summation in Equation 3-3 is over multiple short time periods (each period typically 1 h) within the latest few hours during which precipitation occurred. Hereafter, we refer to a pair of values G_i and R_i as a "gauge–radar pair" (GRP). The bias correction is applied through

$$P_{mfc} = \beta_{mf} R_0 \tag{3-4}$$

where R_0 = the original radar precipitation estimate and P_{mfc} = the radar-estimated precipitation corrected for mean-field bias. The correction in Equation 3-3 is applied at all points within the radar umbrella.

It should be noted that in the discussion that follows, it is assumed that for all pairs, both G and R are nonzero or above the minimum precipitation detection threshold. This condition is designed to account for intermittency in the respective precipitation fields because radar and gauge detections do not match precisely, particularly in conditions of light precipitation (Smith and Krajewski 1991, Seo 1998). The formulation in Equation 3-3 also optimizes long-term agreement in precipitation amounts as indicated by the gauge network and by the radar. Although a mean-field bias factor could be estimated by averaging gauge/radar ratios at individual gauge sites, this would not be optimal in a long-term mass balance sense.

Operational practice in the NWS has been to rely on an estimate of B_{mf} based on at least 10 GRPs during the most recent hour or a combination of more than 10 GRPs from a longer time interval. The formulation by Seo et al. (1999) employs a table of biases based on multiple time intervals updated hourly. The "effective" number of GRPs in each estimate is determined such that older pairs receive a fractional weight lower than recent ones. The most recent bias value containing enough "effective" pairs is applied to the latest radar data.

The DPA products are disseminated with a text portion containing current estimates of the mean-field bias for each of several temporal averaging intervals ranging from one hour to multiple days. Bias correction is never applied to the values in the DPA product itself. Bias correction is sometimes applied to DSP products, as indicated by a flag in the metadata portion of the individual product. In practice, it is possible to draw GRPs from as far away as the limit of effective radar coverage, nominally 230 km from the radar. However, bias has a range dependency, and bias factors tend to increase with increasing range because of the detection of snow rather than rain and possibly nondetection of precipitation forming below the radar beam (see Figure 3-3). Therefore, if the radar's estimates need to be used only out to a shorter range, such as 150 km, the best practice is to draw GPRs only from within that range. Accounting for range effects on the bias is one of the best arguments for application of the approach described in Section 3.6.2, namely, local bias correction.

3.6.2 Local Bias Correction

Local bias correction logic is very similar to that of mean-field bias correction as illustrated by Equation 3-3, except that the spatial domain (summation over index i) is limited to a set of gauge points near the geographic point in question, typically 40 km to 60 km; the multiplicative correction factor is then applied only at that point. Therefore, the bias factor varies in space, and with an adequately dense gauge network, it can mitigate some range artifacts in the radar QPE field. An algorithm for local bias correction was presented by Seo and Breidenbach (2002) and implemented operationally in AWIPS. This approach has the advantage of accounting for range effects on the bias, although the bias estimates can be affected by gaps in the gauge network. The disadvantage is that fewer GPRs can be applied to the estimate, and a larger time window might be required to obtain a sufficient number of pairs. The longer time window might incorporate multiple precipitation regimes with different bias factors. However, as shown by Seo and Breidenbach (2002), local correction can improve on mean-field correction with even a rather sparse gauge network.

3.7 APPROACHES TO GAUGE–RADAR OBSERVATION MERGING

The methods described in the previous sections all modify radar estimates toward a closer agreement with gauge reports. Depending on the formulation of the bias interpolation method, the local bias correction can be designed to force the final estimate to agree with gauge reports at their site locations given that the radar estimate is nonzero. A major limitation of the multiplicative bias correction approach is that it cannot correct radar estimates of zero at points where reliable gauges indicate precipitation, for example, where the radar beam is overshooting or blockage causes radar underestimation.

In current NWS operations, two methods for the direct merging of gauge and radar estimates are available, as described by Seo (1998) and by McCormick (1995). That by Seo is referred to as double optimal interpolation (DOE), which can also be applied to rain-gauge data alone. The other, referred to as P3 (or P1 for Process 1 in earlier references), was documented by the USACE and McCormick (1995). The DOE accounts explicitly for spatial intermittency in nonzero precipitation; P3 is designed primarily for fast execution and use in areas with a uniform gauge network and no sharp climatic precipitation gradients (Seo 1998).

The principles for directly merging gridded radar and point gauge estimates can be logically expressed in highly simplified form as

$$PM_j = \sum_n \Lambda_{gn} G_n + \Lambda_r R_j \qquad (3\text{-}5)$$

where
PM_j = the final estimate at position j,
G_n = the gauge report at a site n within some locus near point j,
R_j = the radar estimate at point j,
Λ_{gn} = the statistical weight associated with gauge site n, and
Λ_r = the weight associated with the radar observation.

The weights Λ are selected such that they minimize the error variance among collocated gauge–radar pairs and that they sum to unity. It is assumed that the radar estimates, R_j, have been corrected for mean or local bias (Seo 1998).

The alternative logic referred to as P3 (previously cal_rad and Process 1, McCormick 1995) also directly blends gauge reports with the radar field and can yield nonzero precipitation where the radar estimate is zero. The approach relies on formulation of distinct additive and multiplicative bias factors to be applied to the original radar grid.

$$P3_j = \frac{R_j}{N}\sum_{k=1}^{N} w_k \beta m_k + \frac{1}{N}\sum_{k=1}^{N} w_k \beta a_k$$

$$\beta m_k = \begin{cases} 1, & \text{if } G_k/R_k > \beta_{crit} \\ G_k/R_k, & \text{otherwise.} \end{cases}$$

$$\beta a_k = \begin{cases} G_k - R_k, & \text{if } G_k/R_k > \beta_{crit} \\ 0, & \text{otherwise.} \end{cases} \qquad (3\text{-}6)$$

where
k = the index of the nearest set of N gauge locations closest to grid location j,
β_m and β_a = the multiplicative and additive bias factors, respectively, and
w = a weight inversely proportional to the distance between k and j.

The terms R and G represent radar and gauge estimates, respectively. In the operational formulation, N is defined as 3 for the gauge locations defining a triangle enclosing the point j.

It can be seen that the multiplicative factor at a gauge point is equal to the gauge/radar ratio as long as the radar value is nonzero and the ratio is less than some critical value β_{crit} (typically 2); otherwise, the multiplicative factor is defined as unity. If the multiplicative ratio is defined as nonunity, then the local additive bias is set to zero at that point. If the gauge measurement is nonzero but the radar estimate is zero or much smaller than the gauge value, the additive bias is set to the gauge minus radar difference, and the multiplicative bias is set to unity. The additive and multiplicative biases defined at the gauge sites are then separately interpolated to all grid points and applied to the radar field.

The effects of the introduction of radar data and gauge–radar merging on the spatial structure of rainfall fields are illustrated in Figure 3-5, which shows 1-hour estimates shortly after the landfall of Hurricane Dennis over the southeastern United States on July 10, 2005. A field constructed solely from gauge reports [Figure 3-5(a)] shows the broad outline of the precipitation. Radar estimates without and with mean-field bias adjustment [Figures 3-5(b) and 3-5(c)] were obtained by mosaicking DPA products from multiple radars. They show increased spatial detail, particularly higher accumulations over 75 mm near the remnants of the storm eye over southern Alabama. The bias-adjusted version [Figure 3-5(c)] has a higher maximum in this area. The final gauge–radar merged product is based on the gauge–radar merging logic of Equation 3-5 [Figure 3-5(d)]; it also received manual quality control from the staff of the Southeast River Forecast Center. Although it is similar to the bias-adjusted version in Figure 3-5(c), some radar artifacts are removed, and gauge-detected light rainfall over North and South Carolina was accepted by the analyst and retained in this final version.

3.8 APPLICABILITY OF THE GAUGE–RADAR APPROACHES

As shown by Smith and Krajewski (1991), Seo et al. (1999), and Seo and Breidenbach (2002), the successive refinements of mean-field bias correction, local bias correction, and gauge–radar merging consistently lead to improvements over gauge-only estimates within areas of typical gauge network density. However, questions might arise as to whether radar QPE is needed at all if there is a reliably reporting gauge near (but not necessarily collocated) with the geographic point of interest. Also, the density of the available rain-gauge network can affect the quality of merged gauge–radar products (Habib et al. 2013).

If the problem at hand is to estimate hourly rainfall at a single location, and gauge observations with the necessary temporal resolution are available from a nearby site, it is possible that radar QPE might not improve the gauge estimates. During the warm season, April–September, studies have shown that gauge-to-gauge correlation for hourly rainfall remains above 0.8 for separation distances no larger than 2 km (1.24 mi) (Gebremichael and Krajewski 2004, Berndtsson 1987). Although Young et al. (2000) reported somewhat higher gauge-to-gauge correlations over Oklahoma, they also noted considerable spread in the correlation estimates. Given that gauge–radar correlations for 1-hour data are generally more than 0.7 for quality-controlled radar data and locations less than 150 km from the radar (e.g., Mazari et al. 2013), it appears that radar or gauge–radar estimates will generally be superior to gauge-only estimates for locations more than a few kilometers from the nearest gauge site. In studies conducted over the southern United States

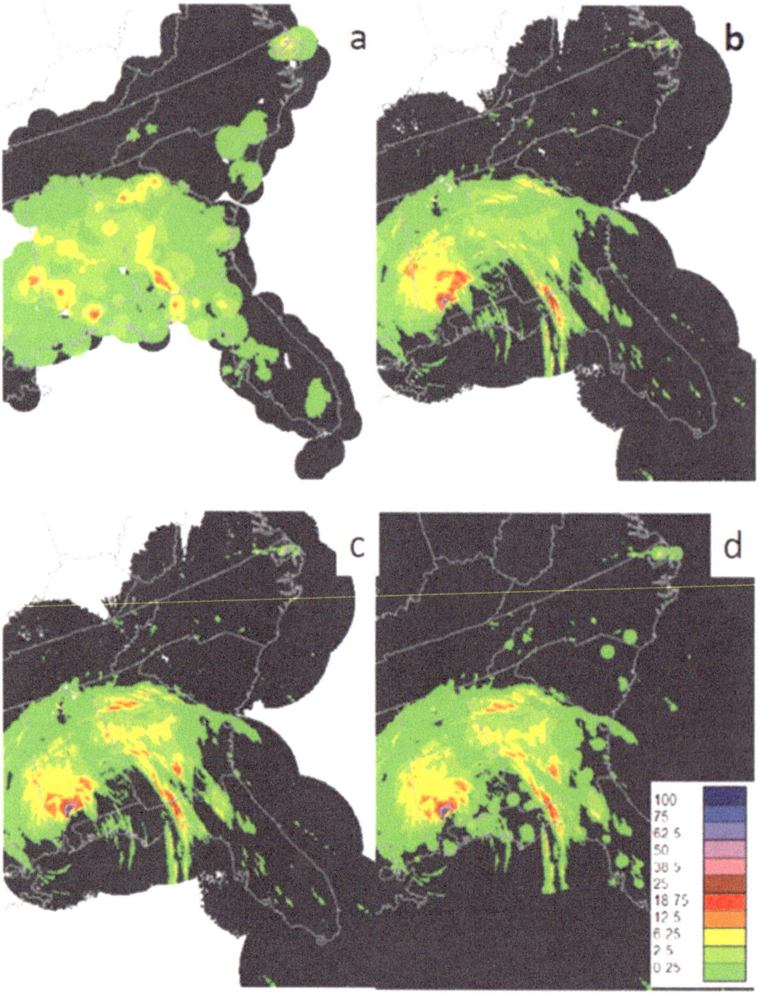

Figure 3-5. Analyses of 1-hour precipitation ending 0000 UTC, July 11, 2005, over the southeastern United States, from (a) rain gauges; (b) radar without bias correction; (c) radar with mean-field bias correction; and (d) gauge-radar merge with manual quality control, from Southeast River Forecast Center. Color scale is in mm.

during the warm season, radar appeared to better detect precipitation than did nearby gauge reports (Wang et al. 2008b) or a gauge-only gridded analysis (Habib et al. 2013).

Although much of the conterminous United States is within 200 km of a WSR-88D unit, terrain blockage and precipitation climatology can render radar unable to reliably detect precipitation at many locations. As shown by Maddox et al. (2002), the area covered by radar beams centered no higher than 2 km above ground level is much smaller than the extent of the 200 km range. Such low-level coverage is necessary to detect raindrops, especially during the cool season.

Users of radar QPE should assess whether radar detects measurable precipitation for approximately the same number of hours as indicated by nearby gauge reports for the period or events under consideration. Further considerations in determining the relative importance of gauge and radar estimates in merging these two data sources, and in designing gauge networks for use in combination with radar data, are documented in Chapters 4, 7, 8, and 9.

3.9 USE OF DAILY PRECIPITATION REPORTS IN COMBINATION WITH RADAR QPE

The discussion to this point has been focused strictly on radar and gauge estimates of 1 h duration. However, much information that further increases the reliability of these estimates can be obtained from daily-reporting gauges. The overall density of the daily-reporting gauge network, both real-time and delayed, is much higher than that of the hourly-reporting network. Moreover, many daily reports are from collecting gauges, which are less subject to some mechanical error sources than are the tipping bucket units commonly used for hourly observations (Wang et al. 2008a, Habib et al. 2001).

It is possible to use these daily gauge reports in combination with hourly radar estimates to construct a fairly reliable hourly precipitation time history that is unbiased at the daily time scale. In the simplest application, radar or gauge–radar gridded data collocated with a daily gauge site near the point of interest can be used to disaggregate the daily reports to subdaily. This is essentially bias adjustment based on daily precipitation totals. Alternatively, daily gauge reports can be applied in a final bias adjustment of hourly analyses derived in real time by one of the methods described above. That is, the 24 h total of the hourly or other subdaily estimates is constrained to match collocated gauge-derived daily totals and the individual hourly values adjusted accordingly. Such adjustments have been applied retrospectively to operational multisensor analyses that utilized only a limited set of gauge reports owing to operational time constraints (e.g., Zhang et al. 2011b). Both radar and satellite precipitation gridded hourly estimates were applied to time disaggregation of daily precipitation analyses in the construction of the North American Land Data Assimilation (NLDAS) data set (Cosgrove et al. 2003).

An alternative application of daily gauge observations involves adjustment of real-time gauge–radar data based on their long-term statistical relationships to daily gauge-based analyses. The Climatology-Calibrated

Precipitation Analysis (Hou et al. 2013) utilizes regression-based corrections to the real-time Stage IV product, with the corrections being unique to each grid box and season. The approach corrects biases owing to radar range and inhomogeneous site distributions in the real-time gauge network.

3.10 ACCESS TO PRECIPITATION OBSERVATIONS AND ESTIMATES

3.10.1 Products from Individual Radar Sites

Both base radar data (reflectivity, Doppler velocity, and velocity spectrum width) and certain derived products (rainfall, vertically integrated liquid, vertical composite reflectivity) are generated at individual radar sites and ported to several processing hubs within and outside NOAA. Real-time graphical images are posted through the web portal http://www.weather.gov/Radar and by some commercial vendors. Reflectivity data from the WSR-88D and TDWR networks are available in binary file format in real time from some commercial vendors and from the NWS through the Radar Product Central Collection Dissemination Service (RPCCDS, NWS 2009). Data from any one site are packaged based on radar antenna elevation angle. Both reduced-precision (4-bit) and full-precision (8-bit) product files are available. The DHR and DSP (azimuth-range) and DPA (Cartesian grid) products are available through this dissemination system. Archived radar base data (Level II) and derived radar products including rainfall estimates (Level III) are available from NCEI through the Archive Information Request System (AIRS, NCEI 2017). Software tools for interactive visualization and interpretation of these data include the Weather and Climate Toolkit (WCT, NCEI 2016).

3.10.2 Gridded Multi-Radar and Multi-Sensor Products

Automated and manually adjusted high-resolution precipitation products are prepared by the National Centers for Environmental Prediction (NCEP) and by RFCs. National-scale 1 h, 4 km grids for the conterminous United States are routinely produced. Gridded radar-only estimates referred to as Stage II (Lin and Mitchell 2005) are produced by NCEP for use in operational land-surface models. Gauge–radar grids with manual quality control and adjustment are prepared at RFCs for hydrologic model input and are composited to national-scale product referred to as Stage IV. The Stage IV operation is carried out within NCEP (Lin and Mitchell 2005). These grids are available in graphical and GIS-compatible formats through the web portal http://water.weather.gov/precip/.

Both Stage II and Stage IV estimates can be obtained on a delayed basis from the University Corporation for Atmospheric Research (UCAR 2018a,b).

End users might wish to utilize one of these gridded precipitation products in conjunction with separately collected or newly available gauge reports, following any of the methods outlined in Sections 3.5 and 3.6. Stage II and Stage IV products are formatted as GRIdded Binary (GRIB1) data. Software tools for decoding of this format include wgrib (NCEP 2005) and degrib (Meteorological Development Laboratory 2010).

Real-time precipitation products from the MRMS are currently available via web interface from NCEP (http://mrms.ncep.noaa.gov/data/).

3.10.3 Rain-Gauge Reports

Archives of observations from rain-gauge sites reporting in near-real time through the Hydrometeorological Automated Data System (HADS, NCEP 2018), typically in 1 h intervals, are available from UCAR through a web interface. Additional reports from hourly and sub-hourly gauge sites not reporting in real time are available from the NCEI, again through a web interface. A processing fee might be charged to nongovernmental and non-university organizations. Daily reports (24 h totals) are also available from these sources.

3.11 CONCLUSIONS

We have presented a summary of the challenges to and advantages of applying radar precipitation estimates, with and without accompanying rain-gauge data. Despite the rather large error distributions associated with radar estimates, they are superior to gauge-only estimates in many situations where collocated gauge data are not available. Ongoing improvements to radar estimates through the implementation of dual-polarization technology (Giangrande and Ryzhkov 2008) and continued development of single-polarization precipitation and quality control algorithms, such as those described by Zhang et al. (2011a) promise to make radar a standard precipitation reference shortly. These advances, along with the general availability of estimates in digital form through web protocols, should encourage all practitioners needing remote precipitation estimates to consider their use.

REFERENCES

Anagnostou, E., and W. Krajewski. 1998. "Calibration of the WSR-88D precipitation processing subsystem." *Weather Forecast.* 13(2), 396–406.

Bellon, A., G.-W. Lee, and I. Zawadzki. 2005. "Error statistics of VPR corrections in stratiform precipitation." *J. Appl. Meteor.* 44, 998–1015.

Berndtsson, R. 1987. "Spatial and temporal variability of rainfall and potential evaporation in Tunisia." *IAHS Publ.* 168, 91–100.

Brandes, E. A. 1975. "Optimizing rainfall estimates with the aid of radar." *J. Appl. Meteor.* 14, 1339–1345.

Chilson, P., W. F. Frick, J. F. Kelly, K. W. Howard, R. P. Larkin, R. H. Diehl, et al. 2012. "Partly cloudy with a chance of migration: Weather, radars, and aeroecology." *Bull. Amer. Meteor. Soc.* 93(11), 669–686.

Ciach, G., W. Krajewski, and G. Villarini, G. 2007. "Product-error-driven uncertainty model for probabilistic quantitative precipitation estimation with NEXRAD data." *J. Hydrometeorol.* 8, 1325–1347.

Cosgrove, B., D. Lohmann, K. E. Mitchell, P. R. Houser, E. F. Wood, J. C. Schaake, et al. 2003. "Real-time and retrospective forcing in the North American Land Data Assimilation System (NLDAS) project." *J. Geophys. Res.* 108, 8842.

Doviak, R., and D. Zrnić. 2006. *Doppler radar and weather observations.* 2nd Ed. Mineola, NY: Dover Publications, 592.

Fulton, R., J. Breidenbach, D. -J. Seo, D. Miller, and T. O'Bannon. 1998. "The WSR-88D rainfall algorithm." *Weather Forecast.* 13(2), 377–395.

Fulton, R., F. Ding, and D. Miller. 2003. "Truncation errors in historical WSR-88D rainfall products." *Proc., 31st Conf. Radar Meteorology.* Boston: American Meteorological Society, 6–12.

Gebremichael, M., and W. Krajewski. 2004. "Assessment of the statistical characterization of small-scale rainfall variability from radar: Analysis of TRMM ground validation datasets." *J. Appl. Meteor.* 43, 1180–1199.

Giangrande, S., and A. Ryzhkov. 2008. "Estimation of rainfall based on the results of polarimetric echo classification." *J. Appl. Meteor. Climatol.* 47, 2445–2462.

Habib, E., W. F. Krajewski, and A. Kruger. 2001. Sampling errors of tipping-bucket rain gauge measurements. *J. Hydrol. Eng.* 6, 159–166.

Habib, E., L. Qin, D. -J. Seo, G. J. Ciach, and B. R. Nelson. 2013. "Independent assessment of incremental complexity in NWS multisensor precipitation estimator algorithms." *J. Hydrol. Eng.* 18, 143–155.

Hou, D., M. Charles, Y. Luo, Z. Toth, Y. Zhue, R. Krzysztofowicz, et al. 2013. "Climatology-calibrated precipitation analysis at fine scales: Statistical adjustment of Stage IV towards CPC gauge-based analysis." *J. Hydrometorol.* 15(6), 2542–2557.

Kitzmiller, D., D. Miller, R. Fulton, and F. Ding. 2013. "Radar and multisensor precipitation estimation techniques in National Weather Service hydrologic operations." *J. Hydrol. Eng.* 18, 133–142.

Krajewski, W. F., and J. A. Smith. 2002. Radar hydrology: Rainfall estimation. *Adv. Water Res.* 25(8-12), 1387–1394.

Lakshmanan, V., T. Smith, G. Stumpf, and K. Hondl. 2007. "The warning decision support system—integrated information." *Weather Forecast.* 22(3), 596–612.

Lakshmanan, V., J. Zhang, and K. Howard. 2010. "A technique to censor biological echoes in weather radar images." *J. Appl. Meteor. Climatol.* 49, 453–462.

Lin, Y., and K. Mitchell. 2005. "The NCEP stage II/IV hourly precipitation analyses: Development and applications." *Proc., 19th Conf. Hydrology*. Boston: American Meteorological Society.

Maddox, R., J. Zhang, J. Gourley, and K. Howard. 2002. "Weather radar coverage over the contiguous United States." *Weather Forecast.* 17(4), 927–934.

Marshall, J., and W. Palmer. 1948. "The distribution of raindrops with size." *J. Meteorol.* 5, 165–166.

Mazari, N., H. Xie, J. Zeitler, and H. Sharif. 2013. "Validation of the NEXRAD DSP product with a dense rain gauge network." *J. Hydrol. Eng.* 18, 156–167.

McCormick, B. 1995. "RainView (cal_rad subsection)." *U.S. Geological 845 Survey*. Tulsa, OK: Tulsa District Office.

Meteorological Development Laboratory. 2010. "Degrib (NDFD GRIB2 Decoder)." Accessed April 2013. http://www.nws.noaa.gov/mdl/degrib.

NCEI (National Center for Environmental Information). 2017. "Archive information request system (AIRS)." Accessed October 2017. http://www.ncdc.noaa.gov/has/HAS.DsSelect.

NCEI. 2016. "NOAA's weather and climate toolkit." Accessed October 2017. http://www.ncdc.noaa.gov/wct.

NCEP (National Centers for Environmental Prediction). 2005. "wgrib." Accessed April 2013. http://www.cpc.ncep.noaa.gov/products/wesley/wgrib.html.

NCEP. 2018. "Hydrometeorological Automated Data System." Accessed August 2018. http://hads.ncep.noaa.gov/index.shtml.

Park, H.-S., A. Ryzhkov, D. Zrnić, and K.-E. Kim. 2009. "The hydrometeor classification algorithm for the polarimetric WSR-88D: Description and application to an MCS." *Weather Forecast.* 24(3), 730–748.

Seo, D.-J. 1998. "Real-time estimation of rainfall fields using radar rainfall and rain gage data." *J. Hydrol.* 208, 37–52.

Seo, D.-J, J. Breidenbach, and E. Johnson. 1999. "Real-time estimation of mean-field bias in radar rainfall data." *J. Hydrol.* 223, 131–147.

Seo, D.-J., J. Breidenbach, R. Fulton, D. Miller, and T. O'Bannon. 2000. "Real-time adjustment of range-dependent biases in WSR-88D rainfall estimates due to nonuniform vertical profile of reflectivity." *J. Hydrometeor.* 1, 222–240.

Seo, D.-J., and J. Breidenbach. 2002. "Real-time correction of spatially nonuniform bias in radar rainfall data using rain gauge measurements." *J. Hydrometeor.* 3, 93–111.

Smith, J. A., and W. Krajewski. 1991. "Estimation of the mean-field bias of radar rainfall estimates." *J. Appl. Meteor.* 30, 397–412.

Smith, J. A., and W. Krajewski. 1994. "Estimation of parameters for the NEXRAD rainfall algorithms." *Final Rep.*, Hydrology Laboratory, W/OHD, NWS/NOAA, Silver Spring, MD, 96 pp.

UCAR (University Corporation for Atmospheric Research). 2018a. "NCEP/EMC U.S. Gridded Radar-Estimated Precipitation w/ no bias removal (4km)." Accessed August 2018. http://data.eol.ucar.edu/cgi-bin/codiac/fgr_form/id=21.090.

UCAR. 2018b. "GCIP/EOP Surface: Precipitation NCEP/EMC 4KM Gridded Data (GRIB) Stage IV Data." Accessed August 2018. http://data.eol.ucar.edu/cgi-bin/codiac/fgr_form/id=21.093.

Wang, J., B. Fisher, and D. Wolff. 2008a. "Estimating rain rates from tipping-bucket rain gauge measurements." *J. Atmos. Oceanic Technol.* 25, 43–56.

Wang, X., H. Xie, H. Sharif, and J. Zeitler. 2008b. "Validating NEXRAD MPE and stage III precipitation products for uniform rainfall on the Upper Guadalupe river basin of the Texas Hill Country." *J. Hydrol.* 348(1–2), 73–86.

Wilson, J. 1970. "Integration of radar and rain–gauge data for improved rainfall measurement." *J. Appl. Meteor.* 9, 489–497.

Xu, X., K. Howard, and J. Zhang. 2008. "An automated radar technique for the identification of warm rain processes." *J. Hydrometeorol.* 9(5), 885–902.

Young, C., A. A. Bradley, W. F. Krajewski, A. Kruger, and M. Morrissey. 2000. "Evaluating NEXRAD multisensor precipitation estimates for operational hydrologic forecasting." *J. Hydrometeorol.* 1, 241–254.

Zhang, J., and Y. Qi. 2010. "A real-time algorithm for the correction of bright band effects in radar-derived precipitation estimation." *J. Hydrometeorol.* 11(5), 1157–1171.

Zhang, J., K. Howard, C. Langston, S. Vasiloff, B. Kaney, A. Arthur, et al. 2011a. "National Mosaic and Multi-Sensor QPE (NMQ) System: Description, results, and future plans." *Bull. Amer. Meteor. Soc.* 92(10), 1321–1338.

Zhang, Y., S. Reed, and D. Kitzmiller. 2011b. "Effects of retrospective gauge-based readjustment of multi-sensor precipitation estimates on hydrologic simulations." *J. Hydrometeor.* 12(6), 1414–1431.

Zhang, J., K. Howard, C. Langston, B. Kaney, Y. Qi, L. Tang, et al. 2016. "Multi-Radar Multi-Sensor (MRMS) quantitative precipitation estimation: Initial operating capabilities." *Bull. Amer. Meteor. Soc.* 97, 621–637.

CHAPTER 4

EVALUATION AND IMPROVEMENT OF RADAR RAINFALL DATA

4.1 RAINFALL MEASUREMENT METHODS, ERRORS, AND ACCURACY

Radar rainfall estimates and rain-gauge measurements are prone to systematic and random errors. The field of metrology (the science of measurement) provides an exhaustive treatment of different types of errors and their manifestations in different measurement settings. Systematic errors that are mainly due to measuring instruments (instrumental systematic errors, Rabinovich 2005). These errors can be corrected but cannot be eliminated completely. In some situations, systematic errors are qualified as personal systematic errors (Rabinovich 2005), and they are mainly due to the individual characteristics of the observer. Systematic errors at rain gauges due to wind undercatch, splashing, evaporation, and wetting losses are also common. Random errors are mainly due to imprecision in measurement and unknown wind fields responsible for losses (Vieux 2001), irregularities of topography, and micro-climatological variations around the gauge site and inadequate network density to account for the natural spatial variability of rainfall [WMO (World Meteorological Organization) 2009].

4.2 RAIN-GAUGE AND RADAR-BASED MEASUREMENTS

Ground-based measurements refer to the conventional and direct ways of measuring rainfall, and these are obtained by a network of rain gauges. Radar and satellite-based measurements are more recent methods for estimating rainfall. Sene (2009) documents the available satellite-based rainfall estimates derived from different algorithms.

According to Servuk and Klemm (1989), there are more than 150,000 nonrecording rain gauges throughout the world. There are also several different types of gauges that differ in catch area, height, and material used and other aspects. Servuk and Klemm (1989) and Strangeways (2007, 2003) indicate that there are more than 50 types of manually read gauges in the world. These estimates may not reflect current figures, and the variations in the types are not easy to document. However, these numbers and types of gauges show the diversity in the configuration of rain gauges around the world. Recording gauges measure rainfall at discrete time intervals or the timing of the increments and thus provide the time distribution of rainfall for a given storm event or within a time frame of consideration. Typically, recording gauges are electronically controlled, and the information about rainfall measurements is transmitted through telemetric devices. Examples of these types of gauges include tipping bucket, float, and weighing bucket gauges. Nonrecording gauges measure rainfall without providing a time distribution of rainfall. The nonrecording rain gauge, consisting of a reservoir with a funnel, is conceptually the simplest device for measuring rainfall. Because no automatic recording devices are attached, rainfall measurement readings have to be taken manually at regular time intervals, usually daily.

There are several types of weather radars that are in use throughout the world. They are sometimes classified based on the frequency and wavelength characteristics. An exhaustive discussion of radar-based rainfall measurements can be obtained from reference books by Meischner (2005), Strangeways (2007), and Raghavan (2003). Meischner (2005) provides a comprehensive assessment of radar-based rainfall networks, radar rainfall estimation, the use of radar data for hydrometeorological applications, and advances in the radar measurement of rainfall. The WSR-88D methodology and algorithms for radar-based rainfall estimation are described in detail by Fulton et al. (1998). The reader is referred to Chapter 2 of this manual on radar-based rainfall estimation and the works of Collier (1989) and Vieux (2001) for initial developments of radar-based rainfall networks and their evolution. According to Wilson and Brandes (1979), better estimates of rainfall are obtained from bias-adjusted radar rainfall data than can be achieved from either rain-gauge networks or radar rainfall data alone.

4.3 IMPROVING RADAR-BASED ESTIMATION: OPTIMAL *Z–R* RELATIONSHIPS

The use of *Z–R* relationships for estimation of rainfall and commonly used *Z–R* relationships are discussed in Chapter 3. Teegavarapu and Pathak (2012) suggest that optimal *Z–R* relationships can be developed to improve the radar-based precipitation estimation. As precipitation patterns in terms of spatial and temporal variability evolve over time, *Z–R* relationships that are time-dependent and those that vary with the time can be optimized. A varying "temporal window" can be selected for optimizing different *Z–R* relationships over time as the storm event unfolds. Temporal window in this context refers to a specific length of time frame. Optimal coefficient and exponent values are achieved by using optimization formulations. The temporal window size for optimization increases as the storm event unfolds, providing more information about the storm in optimizing the power relationship. The near-time optimal *Z–R* relationship is more suitable for a storm type that is not changing over a shorter duration of time (stationarity is assumed). Similarly, multiple *Z–R* relationships using a constant temporal window and seasonal *Z–R* relationships (Teegavarapu and Pathak 2012) can be developed. A select number of storm events can be used for development of optimized *Z–R* relationships using the probability distributions of reflectivity values. The implementation of this approach in real time is difficult as distributions are not known a priori. The methods for optimizing *Z–R* relationships have shown improvements over other methods. Readers are referred to Chapter 7 and the works of Ferreira et al. (2009), Teegavarapu (2009), and Teegavarapu and Pathak (2012) for details of these methods and their applications.

Incorrectly specified *Z–R* relationships are less of a problem for nonreal-time applications than for real-time data. In principle, this is true in real time, but bias estimation is, of course, more problematic in real time. Analysis of uncertainties in *Z–R* relationships for different storm events and selection of optimal exponents and coefficients requires selection of a number of rain gauges within a region. Selection of observations at rain-gauge stations for optimization of analysis of *Z–R* parameters can be based on several criteria. These criteria include (1) rain-gauge stations reporting positive rainfall, (2) a threshold correlation coefficient estimated based on application of a standard *Z–R* (e.g., $300\,R^{1.4}$) relationship to rain-gauge and radar-based observations, (3) stations located within a specific distance from the radar site, and (4) rain-gauge stations recording positive rainfall in a temporal window selected.

Each event (any temporal frame) with sub-temporal events is analyzed visually to select a window or multiple windows based on the occurrence of positive rainfall. Temporal windows in which a large number of rain gauges report rainfall can be selected for obtaining optimal relationships. Two windows are selected: one for optimizing the parameters and another for testing the optimized parameters values via a number of performance measures.

The following are key issues that need to be considered when optimal or revised *Z–R* relationships are developed (Teegavarapu 2012) and used for estimation of radar-based rainfall:

- Functional forms for the *Z–R* relationship different from the traditional or standard *Z–R* power relationship for a given temporal window (e.g., 15 min interval, day, month, or season) can be used for a region for obtaining improved rainfall estimates.
- Coefficient and exponent parameter values of power relationship (i.e., $Z = a R^b$) can be obtained via optimization considering the realized storm events. The minimum and maximum values of coefficients and exponents need not be fixed a priori for optimization formulations.
- Improved estimates of rainfall can be realized when reflectivity and rainfall rate information from temporal window preceding the window in which the optimal relationship of *Z–R* is applied, is available. In real time, improvement in the estimates will depend on the amount of information available in the elapsed time interval with observed rainfall at rain gauges.
- Accurate gap and error-free continuous rain-gauge observations are critical to the success of the *Z–R* optimal relationships.
- Storm type information can be incorporated within the optimization or selection of *Z–R* relationships by analyzing the raw reflectivity values.
- Number of rain gauges used for developing optimal *Z–R* relationships depends on the performance criteria used in the process of selecting the number. However, to obtain a robust estimate of coefficients and exponents of *Z–R* relationships, all available error-free rain gauges should be used.
- Rain gauges can be selected based on a performance criterion such as correlation coefficient between radar-based rainfall estimates using standard *Z–R* relationships and co-located rain-gauge estimates. Rain gauges with high correlation with preliminary radar-based estimates (based on the standard *Z–R* relationship) in deriving the optimal *Z–R* relationship can be used.
- Constant and varying window approaches are applicable for optimization of *Z–R* relationships in real time provided an optimization solver is linked to a radar data processing tool and the execution time of the solver in obtaining solutions is within the period of interest.
- Average optimal coefficient and exponent values for specific seasons can be recommended for further evaluation and applicability in real time.
- Probability distribution of raw reflectivity values from the past storms events can be used to obtain optimal coefficient and exponent values for similar events in the future.
- Optimal relationships are applicable in real time and can also be adopted for post-processing of the radar data. The seasonal average

optimal coefficient and exponent values obtained from this study can be used as starting values for rainfall estimation before bias corrections are applied.

4.4 CONCLUSIONS

The chapter presented a brief summary of rainfall measurements using ground and radar-based measurements. The focus of this chapter is on evaluation and improvement of radar-based rainfall measurements using optimizing functional forms of reflectivity–rainfall relationships.

REFERENCES

Collier, C. G. 1989. *Applications of weather radar systems: A guide to uses of radar data in meteorology and hydrology*. Hoboken, NJ: Wiley.

Ferreira, A., R. S. V. Teegavarapu, and C. Pathak. 2009. "Evaluation of optimal reflectivity-rainfall (Z-R) relationships for improved precipitation estimates." *EOS Abstract # H31D-0805*. Washington, DC: American Geophysical Union.

Fulton, R., J. Breidenbach, D.-J. Seo, D. Miller, and T. O'Bannon. 1998. "The WSR-88D rainfall algorithm." *Weather Forecast*. 13(2), 377–395.

Meischner, P. 2005. *Weather Radar: Principles and advanced applications*. New York: Springer.

Rabinovich, G. S. 2005. *Measurement errors and uncertainties: Theory and practice*. New York: Springer.

Raghavan, S. 2003. *Radar meteorology*. Dordrecht, the Netherlands: Springer. Atmospheric and Oceanographic Sciences Library.

Sene, K. 2009. *Hydrometeorology: Forecasting and applications*. New York: Springer.

Servuk, B., and S. Klemm. 1989. "Types of standard precipitation gages." *WMO/IAHS/ETH Int. Workshop on Precipitation Measurement*. Geneva: World Meteorological Organization.

Strangeways, I. 2003. *Measuring the natural environment*. 2nd Ed. Cambridge, UK: Cambridge Univ. Press.

Strangeways, I. 2007. *Precipitation: theory, measurement and distribution*. Cambridge, UK: Cambridge Univ. Press.

Teegavarapu, R. S. V. 2012. "Assessment and evaluation of NEXRAD-based rainfall estimates in South Florida Water Management District." *SFWMD Rep.*, West Palm Beach, FL: SFWMD.

Teegavarapu, R. S. V., and C. Pathak. 2012. "Development of optimal Z-R relationships." *Weather Radar Hydrol*. London: International Association of Hydrological Sciences, 75–80.

Vieux, B. E. 2001. *Distributed hydrologic modeling using GIS*. Dordrecht, the Netherlands: Kluwer Academic Water Science and Technology Library.

Wilson, J., and E. Brandes. 1979. "Radar measurement of rainfall: A summary." *Bull. Amer. Meteor. Soc.* 60, 1048–1058.

WMO (World Meteorological Organization). 2009. *Guide to hydrological practices: Volume II, Management of water resources and application to hydrological practices*. WMO No. 168, Geneva: World Meteorological Organization.

CHAPTER 5

USE OF RADAR RAINFALL DATA IN HYDROLOGIC MODELING

5.1 DATA REQUIREMENTS FOR HYDROLOGIC MODELING AND DESIGN

Accurate estimation of volumetric watershed outflow is often the key milestone or objective in hydrologic modeling and design applications. Estimates of volumetric outflow in terms of peak flow, total flow volume, or the time history of volumetric flow rates are produced by a hydrologic model representing watershed processes. These models transform incoming rainfall (i.e., volumetric inflow estimates) into volumetric outflow (see Figure 5-1).

For this discussion, volumetric inflow is defined as the time rate of rainfall volume falling on a target watershed. Rainfall volume is computed as the product of the watershed area and rainfall depth. Watershed area is easily determined precisely by standard methods such as topographic survey, planimetering watershed maps, or GIS tools. However, rainfall depth is determined with much less certainty. In the modeling context, the estimated rainfall surface is critical to hydrologic model performance. Natural watersheds respond to the actual spatial and temporal distribution of rainfall to produce an observed outflow hydrograph. Hydrologic models respond to the estimated rainfall distribution provided as input to produce a simulated hydrograph. The greater the difference between the naturally occurring rainfall surface and the estimated rainfall surface, the greater the hydrologic modeling problem. In these cases, hydrologic models tend to be calibrated to overcome rainfall estimation bias rather than to create a consistent representation of watershed processes.

Therefore, the objective of rainfall estimation for hydrologic modeling and design is to estimate a temporally varying rainfall surface that represents

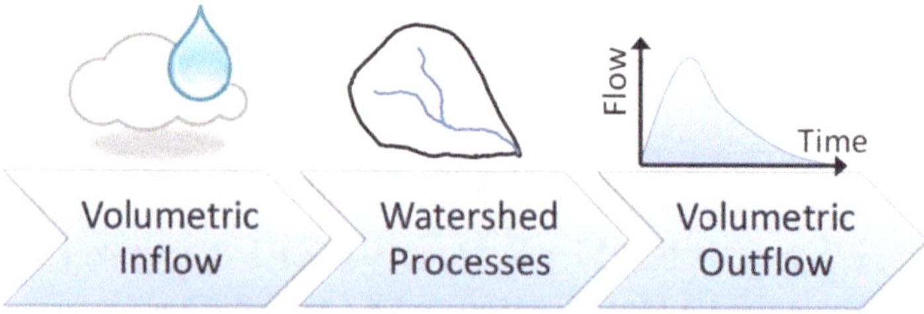

Figure 5-1. Rainfall transformation process.

the actual rainfall surface as accurately as possible. Radar rainfall estimation offers great potential to provide accurate rainfall surfaces because radars can (1) completely cover entire watersheds, (2) provide high spatial resolution, and (3) provide high temporal resolution.

5.2 RADAR-BASED RAINFALL DATA FOR HYDROLOGIC MODELING

5.2.1 Hydrologic Model Calibration

When calibrating hydrologic models, it is important to use the same type of data in calibration as is used in operation. No matter what type of rainfall data are used (i.e., gauge or radar), the calibration process will engrain biases into the model parameters because of differences in the way gauge networks and radars represent rainfall. Therefore, if modelers intend to use GARR data in hydrologic analysis, it is good practice to calibrate the hydrologic model using GARR data.

For discussion purposes here, hydrologic models are grouped into two categories: lumped and gridded. A lumped hydrologic model combines all hydrologic processes within a single watershed or subwatershed as a single element. A gridded model considers hydrologic processes within a rectilinear grid cell. Grid cells are typically on the scale of 1, 4, or 16 km^2 (0.39, 1.54, or 6.18 mi^2). A gridded model normally represents a watershed or subwatershed in much greater detail than a lumped model of the same watershed.

GARR data are used in both lumped and gridded hydrologic models. In lumped models, the gridded GARR data are combined into watershed or subwatershed averages. GIS tools are used to "intersect" the rainfall grid with watershed boundaries to produce watershed average rainfall values at each time step. The time series of watershed averages are input to lumped models exactly like traditional gauge-averaged values. With gridded hydrologic models, the model grid is "intersected" with the rainfall grid to produce an average GARR estimate for each hydrologic model grid cell. Hydrologic model calibration then proceeds traditionally. Hydrologic model calibration and performance is a question that has been investigated by several researchers in recent years. Vieux et al. (2005) demonstrated the successful application of GARR data to a physics-based hydrologic model with a surface resolution of the same scale as the radar data. This study demonstrated the importance of matching the resolution of the input data (radar rainfall estimates) with the resolution of the modeling process.

Kim and Yoo (2014) investigated the response of hydrologic model calibration to replacing gauge-based rainfall estimates with GARR estimates. The study found that GARR estimates were a strong alternative to gauge-based rainfall data because of their spatiotemporal coverage and rarely missing records. Harader et al. (2012) examined error translation between radar rainfall estimates and streamflow simulations. They demonstrated the ability to improve flood forecasting results by correcting radar rainfall estimates using a Kalman filter algorithm. Zhu et al. (2013) also studied the error propagation from radar rainfall estimates to hydrologic models. Most interestingly, they assessed the different results derived from three hydrologic models of varying complexity.

The growing body of successful experience using radar rainfall estimates in hydrologic modeling strongly supports the inherent value of radar rainfall estimation, principally through increased spatial and temporal resolution. Continued research is needed to improve the understanding of the nature of errors propagating from radar rainfall estimation through the stream flow simulation process and how these errors differ from those generated by traditional gauge-only approaches.

5.2.2 Data for Hydrologic Design

The NWS WSR Doppler 88D radars were first installed in the early to mid-1990s. Thus, GARR records are today limited to less than 25 years. Twenty-five years of record is considered too short so support traditional depth–duration–frequency studies. However, the growing national database of radar rainfall estimates represents the first opportunity for hydrologists to conduct in-depth studies of the geometric properties of rainfall events. Curtis et al. (2011), Gill (2005), Hoblit et al. (2004), and Durrans et al. (2002) reported on first efforts exploring storm geometries mined from historical WSR Doppler 88D radar data. In particular, methodologies were explored using GARR radar rainfall data to develop DARFs—a parameter commonly used in the hydrologic design. The results are promising and show the potential for using historical GARR estimates to improve nearly all hydrologic design standards in the United States.

5.3 CONCLUSIONS

Accurate estimation of volumetric watershed outflow is a key objective in hydrologic modeling and design applications. Hydrologic models transform rainfall (i.e., volumetric inflow estimates) into volumetric outflow. The increased spatial resolution of the rainfall field can improve the estimates of watershed inflows, which in turn improve estimates of volumetric outflows. This chapter discussed radar rainfall data formats (i.e., rectilinear grids) for use in hydrologic models. Much work needs to be done, but the early results are promising and show the potential for historical GARR estimates to impact nearly all hydrologic design standards in the United States.

REFERENCES

Curtis, D. C., J. Humphrey, and D. W. Bare. 2011. "Radar-based depth area reduction factors for Colorado." Presented at the American Geophysical Union Fall Meeting, San Francisco, CA, December 5–9.

Durrans, S. R., L. T. Julian, and M. Yekta. 2002. "Estimation of depth-area relationships using radar-rainfall data." *J. Hydrol. Eng.* 7(5).

Gill, T. D. 2005. "Transformation of point rainfall to areal rainfall by estimating areal reduction factors, using radar data, for Texas." Master's thesis, Civil Engineering, Texas A&M Univ. Accessed August 2018. http://hdl.handle.net/1969.1/2420.

Harader, E., V. Borrell-Estupina, S. Ricci, M. Coustau, O. Thual, A. Piacentini, et al. 2012. "Correcting the radar rainfall forcing of a hydrological model with data assimilation: application to flood forecasting in the Lez catchment in Southern France." *Hydrol. Earth Sys. Sci.* 16: 4247–4264.

Hoblit, B., S. Zelinka, C. Castello, and D. Curtis. 2004. "Spatial analysis of storms using GIS." *Proc., 24th Ann. ESRI Int. User Conf.*, ESRI, Redlands, CA.

Kim, J., and C. Yoo. 2014. "Use of a dual Kalman filter for real-time correction of mean field bias." *J. Hydrol.* 519: Part D, 2785–2796.

Vieux, B. E., P. B. Bedient, and E. Mazroi. 2005. "Real-time urban runoff simulation using radar rainfall and physics-based distributed modeling for site-specific forecasts." *Proc., 10th Int. Conf. on Urban Drainage*, Copenhagen, Denmark, August 21–26.

Zhu, D., D. Z. Peng, and I. D. Cluckie. 2013. "Statistical analysis of error propagation from radar rainfall to hydrological models." *Hydrol. Earth Sys. Sci.* 17, 1445–1453.

CHAPTER 6

EXAMPLES IN RADAR RAINFALL DATA, ANALYSES, AND APPLICATIONS

6.1 RADAR RAINFALL ESTIMATION—SOUTH FLORIDA WATER MANAGEMENT DISTRICT

The South Florida Water Management District (SFWMD) has been using radar rainfall data estimation since 2002 (Pathak 2008). Radar rainfall data is used by the SFWMD in making decisions for water management purposes. Five radar stations operated by NWS cover the district: KBYX in Key West, KAMX in Miami, KMLB in Melbourne, KBTW in Tampa, and KJAX in Jacksonville. Although data is also available from private radar installations, the SFWMD exclusively uses NWS sites for its radar rainfall database because of its longevity and reliability. The SFWMD started acquiring radar rainfall data in 2002 in conjunction with three of the other five Florida water management districts; this was subsequently increased to four of the five water management districts. The use of a single vendor for processing radar rainfall data for four water management districts provides an opportunity to eliminate data discontinuities at district boundaries. The use of rain-gauge–adjusted radar rainfall 15 min data (i.e., taken at 15 min intervals) by the SFWMD's Operations Control Center was a major objective of the acquisition.

Near-real-time 15 min rainfall data are provided by using the following process:

1. The SFWMD acquires 15 min radar rainfall accumulations from the NWS, which uses an empirical look-up table to convert reflectivity values to rainfall intensities. Concurrently, 15 min rainfall accumulation data from approximately 152 telemetry rain-gauge sites are sent to the vendor via file transfer protocol (FTP).

2. Vendor adjusts radar rainfall amounts using gauge data algorithms.
3. Vendor places adjusted radar rainfall depths in a flat file and sends via FTP to the SFWMD.
4. The SFWMD checks the flat file for completeness and loads it into an Oracle database.

The process outlined takes 15 to 25 min. Thus, the data are referred to as "near-real-time" with a latency of approximately 20 min. Each file contains 33,773 values, one value for each of the 2 km × 2 km (1.24 mi × 1.24 mi) cells in the grid covering the SFWMD (Huebner et al. 2003). Each 15 min interval file is 366 kb in size. The coverage includes a 35 mi (56 km) area beyond the boundaries of the SFWMD. This provides rainfall information for such areas as the Biscayne and Florida bays and allows operators to assess approaching storms.

Near-real-time data are verified each month, and an end-of-month (EOM) data set of 15 min files is produced. EOM radar rainfall data sets are final data sets and have been through a pre-established QA/QC provided to the SFWMD about 25 to 30 days after the end of the month. The EOM files use data from 81 additional rain gauges that are not available in real time and an algorithm based on the Brandes method to adjust radar rainfall values. An expert from the vendor reviews the results to identify and correct any anomalies or apparent errors. This also provides an ability to identify problematic rain gauges. When the EOM files are received by the SFWMD, the near-real-time data are archived. The current online database contains values from January 1, 1996, to the present. Each cell has a specific time series of rainfall data.

The table structure in the Oracle database complies with the Arc Hydro (Maidment 2002) architecture. Tables using an Arc Hydro schema are produced in Oracle to facilitate corporate implementation of Arc Hydro and provide a common framework for accessing NEXRAD data. Several techniques are used to optimize the database and provide timely access to the data. These include storing only nonzero data values in the database and database partitioning based on the calendar year.

Near-real-time data can be accessed directly using SQL or ODBC drivers. When EOM data sets are available, near-real-time data are replaced. ArcMap-based and text-based applications are available for retrieving gauge-adjusted NEXRAD rainfall data. This web-based user interface allows users to access and aggregate 15 min data for a specified period. The application provides spatially varied (such as rain areas, drainage basins, watersheds) and temporally varied (such as hourly, daily, by event, monthly, and annually) aggregated data sets in tabular and image formats (Pathak 2008).

Skinner et al. (2009) provided a detailed comparison of radar rainfall and rain-gauge measurements in Upper and Lower Kissimmee River basins. Teegavarapu and Pathak (2012) developed an optimal functional relationship between reflectivity (*Z*) and rainfall rate (*R*) relationships. This study evaluated the use of gradient and genetic algorithm-based optimization solvers for optimizing the traditional *Z–R* functional relationships with constants and coefficients for different storm types and seasons for a region in South Florida and their utility in the real-time improvement of rainfall estimates. Rendon et al. (2013) derived a *Z–R* relationship in a sparse rain-gauge network for continuous forecasting in the SFWMD. These relationships were derived through seasonal characterization of gauge and NEXRAD data. Verification of the derived relationships was evaluated using rain gauges withheld for cross-validation. Methods used in forecasting included persistence, seasonal trends, autoregressive, and Kalman filter methods. The results demonstrated that radar-specific relationships exhibited better efficiency than standard relationships and that among the forecasting methods tested, all were advantageous over persistence with varying degrees of accuracy. Seasonal *Z–R* relationships obtained with the seasonal trends (sum of sines) method produced the best results. This method is currently being used in near-real-time radar rainfall estimating at the SFWMD.

6.2 RADAR RAINFALL DATA ANALYSES

Gauge-adjusted radar rainfall data can be treated the same as rain-gauge data for traditional data analyses. However, it is important to preprocess the data before doing data analyses. Preprocessing would vary on a case-by-case basis. For instance, it could include averaging of pixel data for each time step (e.g., 15 min, hourly, or daily) over a spatial domain of interest (i.e., region or watershed).

The following subsections provide some examples of data analyses reported in the literature. One should not consider these listed analyses as a complete set. They are presented here as examples. Many other data analyses can be performed using these data sets.

6.2.1 Rainfall Frequency Analysis

Rainfall frequency analyses are often used to generate "design storms," which are commonly used for flood risk assessment. These design storms are assumed to be uniform in space and standardized in time. Several hydrology texts and reference books provide methods to conduct rainfall frequency data analyses. Pathak (2001) presented frequency analysis of rainfall data in central and south Florida. Wright et al. (2013a, 2013b) presented a method for estimating the frequency of extreme rainfall utilizing radar rainfall data. The method coupled stochastic storm transposition with "storm catalogs" developed from a ten-year radar rainfall record with a

15 min and 1 km^2 (0.39 mi^2) resolution. The storm transposition procedure involved spatial and temporal resampling from these storm catalogs to reconstruct the regional climatology of extreme rainfall. Transposition-based intensity–duration–frequency (IDF) estimates were developed specifically for the selected watershed and did not require simplifying assumptions of storm structure. This procedure was used to reproduce IDF estimates from conventional methods for four urban watersheds in Charlotte, North Carolina. It was demonstrated that extreme rainfall could vary substantially in time and space in ways that cannot be assessed using conventional techniques. Stochastic storm transposition coupled with radar rainfall represented a useful alternative to conventional design storms.

6.2.2 Depth Area Reduction Factors

Curtis (2008) developed a spatial structure of storm cell footprints for Florida. The method was used in computing rainfall DARFs, the sizes and shapes of storms, the distributions of storm speed and direction, and storm orientation. The City of Colorado Springs, Colorado, (Carlton Engineering 2010) developed DARFs for the Fountain Creek watershed in the Colorado Springs area using radar rainfall data sets. A software tool known as TITAN (Thunderstorm Identification, Tracking, Analysis, and Nowcasting) was used to identify and evaluate more than 340,000 15 min storm cells, more than 45,000 1 h storm cells, and approximately 20,000 3 h cells. Storm cell properties were analyzed for eastern Colorado, El Paso County, and the Fountain Creek watershed. Storm-centered DARFs were computed and compared to DARFs published by the NWS that are widely used in hydrologic design. The results showed significant departures from the standard NWS curves. The radar-based storm-centered DARFs developed decay much more sharply than did the standard curves published by the NWS in Technical Paper 29 (US Weather Bureau 1957). It was shown that for a given peak intensity, a population of storm cell sizes exists and that the median storm cell size varies significantly with peak intensity. Thus, the resulting DARFs, which are area dependent, vary with recurrence interval. The completed DARFs suggested sharply lower design rainfall volumes for stormwater design and planning projects.

6.3 OTHER RADAR RAINFALL DATA APPLICATIONS

Radar rainfall data are used in a variety of water resources projects. A small number of applications are presented here as samples. Many other applications are being used by engineers but are not presented.

6.3.1 Near-Real-Time Flood Warning System

Bedient et al. (2000) reported on a flood warning system for the Texas Medical Center (TMC), which focuses on Brays Bayou in southwest Houston, Texas. The system provides the TMC with real-time information on predicted flows in Bray Bayou, along with visual feedback in the form of a real-time BayouCam. Using NEXRAD rainfall and GIS-based analysis, real-time areal-average rainfall over a basin is used to estimate peak flows.

One of the successful applications of the NEXRAD rainfall data was during the Tropical Storm Allison flood in June 2001. It was a highly damaging flood event that impacted Brays Bayou and Houston, causing more than $5 billion in damages in the county and resulting in damage to about 50,000 structures. More than $1.5 billion in damages to the TMC were reported, with major impacts on the local community. Several technical papers have been published on this flood warning system, including Looper and Vieux (2013), Teague et al. (2013), Fang et al. (2011), Fang et al. (2008), and Bedient et al. (2003).

6.3.2 Sewer System Modeling

Urban drainage system design and real-time forecasting for multiscale urban drainage systems are complex. Vieux et al. (2008) developed a distributed hydrologic model that quantified hydrologic prediction uncertainty in urban-scale catchments. The model accurately reproduced the rising limb of the hydrograph well when backwater from Brays Bayou, Houston, Texas, did not impact the outlet hydraulic characteristics. The model also performed well in simulating the peak stage for most events. However, a deviation in the peak simulation stage compared to the observed peak stage was due to unresolved rainfall gradients produced by the NEXRAD radar rainfall. Except for these outliers, the estimated stage prediction accuracy [root mean-square error (RMSE)] was 1.05 ft, and the timing of the peak was predicted with an RMSE of 13.54 min. Results from the simulations showed that the model could make accurate predictions of stage in many cases.

Goormans and Willems (2013) developed a sewer system model using local weather radar data for sewer system operation near Leuven, Belgium. The radar was calibrated using tipping bucket rain-gauge measurements. Various calibration methods were investigated, and a nonlinearly regressed power law function taking a range of the radar into account was found to be the most suitable. Fifty-seven events were simulated using the InfoWorks CS modeling software, and the effect of radar rainfall input on model results was evaluated. The simulation results compared well with the observed flow time series, along with a few large flow deviations in conduits. Storm events showed a higher correlation between measured and simulated flow time series data when using radar rainfall. However, the rain-gauge input data in the model provided slightly better results.

6.3.3 Groundwater Recharge Modeling

Moreno and Vieux (2013) estimated groundwater recharge using a distributed hydrologic model. The groundwater recharge estimate was accomplished using a distributed hydrologic model to estimate components of the hydrologic water balance. The temporal and spatial variability of groundwater recharge were estimated in the eastern outcrop of the Arbuckle–Simpson aquifer, located in Oklahoma. Distributed maps of radar rainfall were bias corrected using rain gauges and used as input to a distributed hydrologic model. Distributed grids of infiltration were combined with evapotranspiration to extend groundwater recharge estimates from three years, when stream flow records existed, to 13 years using radar rainfall data. The results showed that rainfall and runoff were better characterized, with bias-corrected radar rainfall and produced more reliable estimates of runoff and groundwater recharge than those derived from the use of rain-gauge data alone.

6.3.4 Rain-Gauge Network Design

The SFWMD currently operates an extensive network of 279 active rain gauges. A rain-gauge network optimization study was performed (SFWMD 2006) to analyze the existing network and identify areas that have excess or insufficient coverage. Specific recommendations were made for improving the existing rain-gauge network to achieve a consistent level of accuracy across the SFWMD. The optimization study of the existing rain-gauge network using NEXRAD radar rainfall data was performed within an optimal estimation framework that accounted for local rainfall patterns defined by spatial autocorrelation at hourly and daily timescales. The rain-gauge network operates in combination with radar to produce the archival rainfall product for use in operational decision making.

Rainfall measurement accuracy depends on the statistical properties of rainfall and its spatial distribution. A regular array of analysis blocks was used to account for variability in point rainfall processes from the rain-gauge data and the geospatial variability of the radar data. Using the optimal number of gauges per analysis block as a requirement, the study recommended adjusting the existing gauge network with new, relocated, and removed gauges. The study's new network design utilized existing gauges where possible, relocated gauges that were too closely spaced and placed additional gauges such that the optimal gauge number per analysis block was achieved.

The resulting proposed network of 332 (154 proposed and 178 existing) gauges contained the number of total gauges that would be required to achieve an accuracy of a standard error of 0.3 in., which was determined to be reasonable and acceptable to the SFWMD. Of the 154 proposed gauges, 133 would be new and 21 would be relocated. With the addition of new gauges in areas of insufficient coverage, removal of gauges from areas of excess coverage, and relocation of existing gauges, the net increase would be 53 (=332 – 279) gauges. The proposed network with 332 rain gauges would have an average spacing of 12.3 km (7.64 mi), or 150.6 km^2 (58.15 mi^2) per gauge. The result of the study's analysis was a proposed rain-gauge network of variable density that takes into account rainfall statistical properties and the covariant structure of the rainfall field based on radar rainfall data. Implementing these recommendations would provide a network with more varied density and would achieve a consistent level of accuracy across the SFWMD.

6.4 CONCLUSIONS

We have presented examples in radar rainfall data, analyses, and applications in this chapter. Specifically, an example of how radar rainfall data estimation is carried out in the South Florida Management District is presented. The illustrative example also provides all the steps used to generate near-real-time data. Radar rainfall data analyses included for rainfall frequency analysis and DARFs and applications included a near-real-time flood warning system, sewer system modeling, groundwater recharge modeling, and rain-gauge network design are also discussed.

REFERENCES

Bedient, P. B., B. C. Hoblit, D. C. Gladwell, and B. E. Vieux. 2000. "NEXRAD radar for flood prediction in Houston." *J. Hydrol. Eng.* 5(3), 269–277.

Bedient, P. B., A. Holder, J. Benavides, and B. E. Vieux. 2003. "Radar-based flood warning system applied to Tropical Storm Allison." *J. Hydrol. Eng.* 8(6), 308–318.

Carlton Engineering. 2010. *Fountain Creek watershed rainfall characterization study–Final report*. Carlton Engineering, Inc., for City of Colorado Springs, CO.

Curtis, D. C. 2008. "Evaluation of the spatial structure of storms and the development of design storms." *Proc., EWRI World Water and Environmental Resources Cong.*, Reston, VA: ASCE.

Fang, Z., P. Bedient, J. Benevides, and A. Zimmer. 2008. "Enhanced radar-based flood alert system and floodplain map library." *J. Hydrol. Eng.* 13(10), 926–938.

Fang, Z., P. B. Bedient, and B. Buzcu-Guven. 2011. "Long-term performance of a flood alert system and upgrade to FAS3: A Houston, Texas, case study." *J. Hydrol. Eng.* 16(10), 818–828.

Goormans, T., and P. Willems. 2013. "Using local weather radar data for sewer system modeling: Case study in Flanders, Belgium." *J. Hydrol. Eng.* 18(2), 269–278.

Huebner, S., C. S. Pathak, and B. C. Hoblit. 2003. "Development and use of a NEXRAD database for water management in South Florida." *Proc., EWRI World Water and Environmental Resources Cong.*, Reston, VA: ASCE.

Looper, J. P., and B. E. Vieux. 2013. "Distributed hydrologic forecast reliability using next-generation radar." *J. Hydrol. Eng.* 18(2), 260–268.

Maidment, D. R., ed. 2002. *Arc Hydro: GIS for water resources.* Redlands, CA: ESRI Press.

Moreno, M. A., and B. E. Vieux. 2013. "Estimation of spatio-temporally variable groundwater recharge using a rainfall-runoff model." *J. Hydrol. Eng.* 18(2), 237–249.

Pathak, C. S. 2001. "Frequency analysis of daily rainfall maxima for Central and South Florida." *Tech. Pub. EMA No. 390.* West Palm Beach, FL: SFWMD.

Pathak, C. S., ed. 2008. "Hydrologic monitoring network of South Florida Water Management District—Appendix 2-1." South Florida environmental report. Accessed December 15, 2012. https://my.sfwmd.gov/pls/portal/docs/PAGE/PG_GRP_SFWMD_SFER/PORTLET_SFER/TAB2236041/VOLUME1/appendices/v1_app_2-1.pdf.

Rendon, S., B. Vieux, and C. S. Pathak. 2013. "Continuous forecasting and evaluation of derived Z-R relationships in a sparse rain gauge network using NEXRAD." *J. Hydrol. Eng.* 18(2), 175–182.

Skinner, C., F. Bloetscher, and C. S. Pathak. 2009. "Comparison of NEXRAD and rain gauge precipitation measurements in South Florida." *J. Hydrol. Eng.* 14(3), 248–260.

South Florida Water Management District (SFWMD). 2006. "Rain gauge network optimization study–final report." West Palm Beach, FL: Vieux & Associates, Inc.

Teague, A., J. Christian, and P. Bedient. 2013. "Radar rainfall application in distributed hydrologic modeling for Cypress Creek watershed, Texas." *J. Hydrol. Eng.* 18(2), 219–227.

Teegavarapu, R. S., and C. Pathak. 2012. "Development of optimal functional forms of ZR relationships." United Kingdom: IAHS-AISH, 75–80.

US Weather Bureau. 1957. "Rainfall intensity-frequency regime." *Weather Bureau Tech. Paper No. 29.* Washington, DC: Weather Bureau.

Vieux, B. E., J. Imgarten, J. Looper, and P. Bedient. 2008. "Radar-based flood forecasting: Quantifying hydrologic prediction uncertainty in urban-scale catchments for CASA radar deployment." *Proc., EWRI World Water and Environmental Resources Cong.*, Reston, VA: ASCE, 1–10.

Wright, D. B., J. A. Smith, G. Villarini, and M. L. Baeck. 2013a. "Estimating the frequency of extreme rainfall using weather radar and stochastic storm transposition" *J. Hydrol.* 488, 150–165.

Wright, D. B., J. A. Smith, G. Villarini, and M. L. Baeck. 2013b. "Long-term, high-resolution, radar rainfall records for urban hydrology." *J. Am. Water Resour. Assoc.* 50(3), 713–734.

CHAPTER 7

ADVANCED TOPIC: FRAMEWORK FOR BIAS ANALYSIS OF RADAR DATA

7.1 BIAS ANALYSIS METHODS

Radar rainfall estimates derived from the conversion of reflectivity are known to contain systematic errors, bias, and other random errors or artifacts that limit the utility of radar rainfall. Quality control and enhancement of radar rainfall estimates may be accomplished through gauge-adjustment procedures. Improved radar rainfall estimates can be accomplished by many methods. However, all methods are based on the idea of using observations from rain-gauge networks to validate and improve radar rainfall estimates. Krajewski and Smith (2002) ask an important question in their state-of-the-art review of radar rainfall estimation for hydrologic applications: "How good are these estimates?" They suggest validation by a process of determination of the space–time statistical structure of error of radar rainfall products as one of the metrics to evaluate radar data.

Over the last two decades, several studies have evaluated the quality of radar data in different regions of the world and have recommended several bias correction procedures to improve the data. Evaluation of mean-field bias (MFB) and other comparisons of rain-gauge and radar data precede gauge adjustment of radar-based precipitation. Many radar bias correction studies (e.g., Bedient et al. 2013, Young and Brunsell 2008) evaluate two or three indexes for assessment of radar data for correction/adjustment purposes. These indexes mainly rely on rain-gauge (G) to radar (R) value ratios or error measures calculated for different temporal and spatial resolutions. Young and Brunsell (2008) used bias or G/R ratio correlation and detection indexes for evaluation of radar data. Comparison of radar (Stage III NEXRAD) data and co-located rain-gauge data using total difference (i.e., the difference in radar and rain-gauge daily totals), estimation bias, estimation efficiency, and root mean-square deviation (RMSD) was reported by Jayakrishnan et al. (2004). Storm event totals for multiple storms were compared by Lott and Sittel (1996), whereas mean areal precipitation (MAP) values calculated using Thiessen polygons from rain-gauge data and Stage III NEXRAD data were evaluated by Johnson et al. (1999). Comparison of radar-based rainfall products and rain-gauge observations at multiple time scales (hourly, monthly, and daily) was reported by Grassotti et al. (2003). Correlations, MAP, and storm totals were used for comparison in their study.

Steiner et al. (1999) used a ratio of gauge and radar data, RMSE, and accumulated rain and radar data values. Quantitative precipitation forecast skill was used as a measure for evaluating a multiple-radar data set by Xiao and Sun (2007). Spatial distributions of conditional mean and probability of precipitation were evaluated by Young et al. (1999) for evaluation of NEXRAD precipitation estimates in the complex terrain of the northern Appalachian Mountains. Hourly and daily radar estimates and corresponding rain-gauge values were compared using performance measures such as RMSE and correlation by Seo and Breidenbach (2002). Hourly NEXRAD multisensor precipitation estimates and gauge observation-based accumulations were compared by Young et al. (2000). Germann et al. (2006) used bias, false alarm ratio (FAR), the probability of detection (POD), and equitable threat score (ETS) for evaluation of radar data in a mountainous region of the Alps. Scatter plots comparing rain-gauge observations and radar estimates were evaluated by Kitchen et al. (1994) and Vignal et al. (2000). Borga et al. (2000) used mean relative error (MRE), mean absolute error (MAE), and mean basin-average values to compare radar estimates and rain-gauge observations. Brandes (1975) compared radar- and gauge-derived precipitation fields. Scatter plots, correlation coefficient, relative mean bias, and normalized RMSE (NRMSE) were used to compare rain-gauge, radar, and satellite precipitation products by Yilmaz et al. (2005). Scatter plots were used to compare hourly unadjusted and adjusted radar estimates and rain-gauge totals by Seo et al. (1999). Bias, RMSE, correlation, ETS, and FAR are used by Lopez (2011) to evaluate Stage IV radar and gauge data.

Ratios of gauge (G) to radar rainfall (R) were evaluated by Haberlandt (2007) for comparison of geostatistically interpolated hourly precipitation rain-gauge and radar data for extreme rainfall event. Accumulations and scatter plots of rain-gauge observations and radar estimates were compared by Habib and Krajewski (2002) for assessment of bias in radar estimates. Dinku et al. (2002) used MFB to assess the improvements to radar-based rainfall estimation over mountainous terrain. MRE, factional standard error (FSE), and MFB were used for assessment of corrected and uncorrected real-time radar estimates for flood prediction by Borga et al. (2000). Mean-field rain biases were evaluated for WSR-88D data from Tulsa, Oklahoma, by

Anagnostou et al. (1998). MAE was used by Borga (2002) to evaluate basin-average hourly unadjusted and adjusted radar estimates.

An exhaustive review of the literature on bias assessment in radar-based precipitation estimates reveals that several common performance metrics or skills scores were used across a number of studies and regions. Although these measures provided a limited assessment of the quality of radar data in general adequate for the objectives of the studies reported, comprehensive evaluation of pre- or post-corrected radar-based precipitation data quality was rarely carried out in those published studies. A framework for the assessment of radar data is therefore required to ensure that the same can be used for near-real-time operational applications and hydrologic modeling. In a study related to evaluation of different radar-based products at multiple temporal resolutions, Teegavarapu et al. (2015) described a comprehensive framework for the assessment of radar data.

Rain-gauge data can contain errors that could be analyzed for accuracy, temporal consistency, and data gaps. Rainfall measurement accuracy depends on the statistical properties of the rainfall and its spatial distribution. The radar data sets in many situations are gauge adjusted and hence are not completely independent of rain-gauge observations. All the measures and scores are based on the assumption that gauge measurements are ground truth. In reality, gauge error and sampling uncertainty because of the natural spatial variability of rainfall affect the comparative measures, as indicated by Young and Brunsell (2008). Also, according to Teegavarapu et al. (2015), differences in rain-gauge and radar data at 15 min temporal resolution can be attributed mainly to gauge-to-pixel (grid) comparison. In their study, when the temporal resolution of the evaluation was increased from 15 min to 1 h and then to 1 day, improvements in radar data were noted.

Several indexes based on metrics used in hydrologic model evaluation and forecast verification of numerical weather prediction models can be used for assessment of radar rainfall data (Teegavarapu et al. 2015). The indexes can help answer the following major questions:

- What is the overall quality of the radar data compared to rain-gauge data?
- Are characteristics of the radar and rain-gauge estimates statistically similar?
- What is the overall bias?
- What is the error structure, and how does it vary in space and time?
- What is the variability of the bias in space and time?
- What is the skill of the method used for radar-based rainfall estimation in the definition of rain events?
- Do the radar-based rainfall estimates preserve the site (at a rain-gauge location) and regional (a set of rain gauges) statistics?

- Are the rainfall fields generated by radar-based estimates and rain-gauge–based observations similar?
- Can a specific index or a set of indexes help provide directions to improve the radar-based rainfall data?
- How do radar and rain-gauge data quantitatively compare at different spatial and temporal resolutions?
- Is radar data quality appropriate for distributed hydrologic modeling?

It should be noted that rain gauges may also have problems with respect to measuring low/intermittent rainfall rates. Except for the very dense experimental gauge networks, fields from radar and gauge networks will look very different. Although radar provides spatially complete fields, rain gauges generally cannot. Even fields generated using good interpolation methods are very limited, particularly with respect to hydrologic applications where space–time rainfall variability can be extremely important. Assessment of gridded radar data can be carried out using point (rain-gauge) and gridded (interpolated precipitation data from rain-gauge network) data sets. Although many studies compare rain-gauge data (point data) with radar data available at a specific spatial resolution, gridded precipitation data (e.g., Livneh et al. 2013, Maurer et al. 2002), if available, can be used for assessment of radar data. Comparison of point (i.e., rain-gauge) and grid-based (i.e., radar) measurements is the only way to assess radar data when gridded precipitation data based on rain-gauge observations are not available at a specific temporal or spatial resolution. Accurate assessment of radar-based data requires exclusion of suspected anomalous data and erroneous rain-gauge observations.

Bias analyses can be grouped into four categories: (1) temporal, (2) spatial, (3) event-specific, and (4) location-specific. Specific indexes that pertain to each category are presented in this manual. Some of the categories (e.g., spatial, location-specific) can also be analyzed over different time scales. The categories and various levels within each category for evaluations (Teegavarapu et al. 2015) are shown in Figure 7-1.

Evaluation of bias is made by comparing rainfall by collocating rain-gauge and radar data in the fixed tessellations (i.e., grids) of spatial radar-estimated rainfall surfaces. Pooled and stratified samples can be obtained, and they can be evaluated for bias. Pooled sampling involves the pooling of gauge–radar pairs over a specific time or region. Stratified sampling requires the region to be divided into homogeneous or quasihomogeneous areas (Pathak et al. 2009) defined by different criteria. Intraseasonal comparisons will be made at different temporal resolutions (e.g., 15 min, hourly, daily, and monthly). Seasonal comparison of data is made by a grouping of months into different seasons (wet, dry, etc.). Annual evaluations can be used to assess changes in the bias values over the years. Spatial compari-

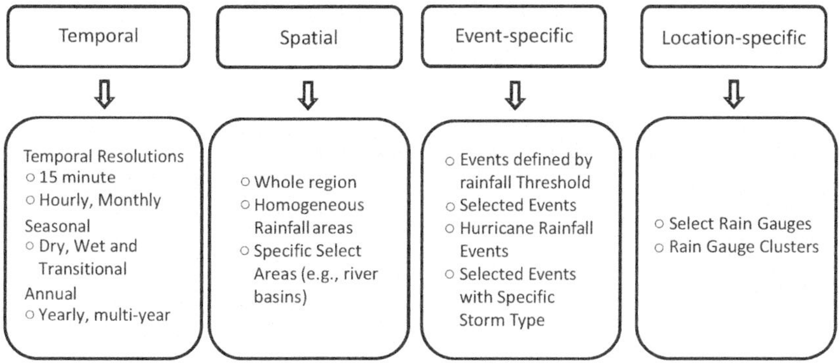

Figure 7-1. Radar data bias evaluation categories.

sons are made using MFB bias values for the whole region or specially identified region (e.g., basin, watershed, meteorologically homogenous area) and spatial variation of biases are evaluated. It should be noted that not all applications require you to conduct such thorough bias analyses as this.

The event-based comparison will involve the use of a few select storm events based on predefined rainfall thresholds and extreme events with high intensity and short duration. Point-based bias evaluation can be addressed by selecting a specific rain gauge or cluster of gauges. Several indexes can be devised to evaluate the biases. Visual indexes will require the development of scatter plots to assess under- or overestimation by radar relative to the selected rain gauges and regions and for different temporal scales (viz. 15 min, day, month, year) along with box plots of two data sets (i.e., rain and radar data). Residual (error) plots can be used to assess the error structure, check for any time dependency of errors, autocorrelation aspects, and heteroskedasticity (Table 7-1).

Accumulated rain and radar plots can be used to evaluate deviations (e.g., divergences) and concurrences. Nonparametric distribution procedures (e.g., kernel density estimates) can be used to evaluate the similarity in distributions of rain and radar data. Combined plots of rain and radar time series can be used to identify the biases and outliers. Error performance measures based on rain-gauge and radar data will need estimation of mean error (ME), MAE, multiplicative bias (MB), RMSE, root mean-square factor (RMSF), linear error in probability space (LEPS), correlation coefficient (CC), anomaly correlation (AC), and Nash–Sutcliffe efficiency coefficient (NSEC) (Nash and Sutcliffe 1970). Skill scores such as the Heidke skill score (HSS), critical success index (CSI), Peirce skill score (PSS), Gilbert skill score (GSS) and odds ratio can also be used. Exhaustive discussion about these skill scores and other forecast verification indexes are provided by Jolliffe and Stephenson (2012).

Table 7-1. Tests for Evaluation of Residuals Obtained from Rain-Gauge and Radar Data

Test or Plot	Evaluation Type	Diagnostics/Inference
Time series plot of residuals (errors)	Time dependency	Random variation and no structured pattern in the variation of residuals with time.
Autocorrelation plot (auto-correlogram)	Check for serial correlation at different temporal lags	Low or negligible correlation at all lags
Probability plot (normal)	Normality of residuals	Linear plot of residuals confirming normality
Histogram	Visual check for normality	Gaussian distribution of errors

The RMSF is the exponent of the RMSE of the logarithm of the data. The logarithmic transformation is mainly performed to smooth the data, reduce the discontinuities, and make the data more robust (CAWCR 2018). Although RMSE can be interpreted as giving a scale to the additive error, the RMSF can be interpreted as giving a scale to the multiplicative error (CAWCR 2018). The LEPS score (Wilks 2011, Jolliffe and Stephenson 2012) is defined as the mean absolute difference between the cumulative probability based on the radar and the cumulative frequency based on the rain-gauge data. The NSEC is generally used in hydrological modeling for assessment of hydrological models. The NSEC can be used to assess the quality of radar-based rainfall estimates compared to the rain-gauge observations. A skill score in forecast verification literature refers to a relative measure of the quality of a forecasting system compared to some benchmark forecasting system. In the context of radar bias analysis, five scores that are relevant to the assessment of radar-based rainfall can be used. The HSS evaluates the data considering the proportion correct (PC) and considering the number of hits due to chance. Hits refer to events (viz., dry or wet) that are correctly defined by radar-based rainfall estimates given rain-gauge observations. The CSI provides information about how well the wet events obtained by radar-based rainfall data correspond to rain-gauge observed wet events. The Pierce skill score addresses the question of radar-based rainfall estimate accuracy in predicting the correct category relative to that of random chance. The GSS, also referred to as the ETS, answers the question, How well did the wet events estimated by radar-based rainfall estimates correspond to the observed wet events (accounting for hits due to chance)? The odds ratio score answers the question, What is the ratio of the odds of a wet event indi-

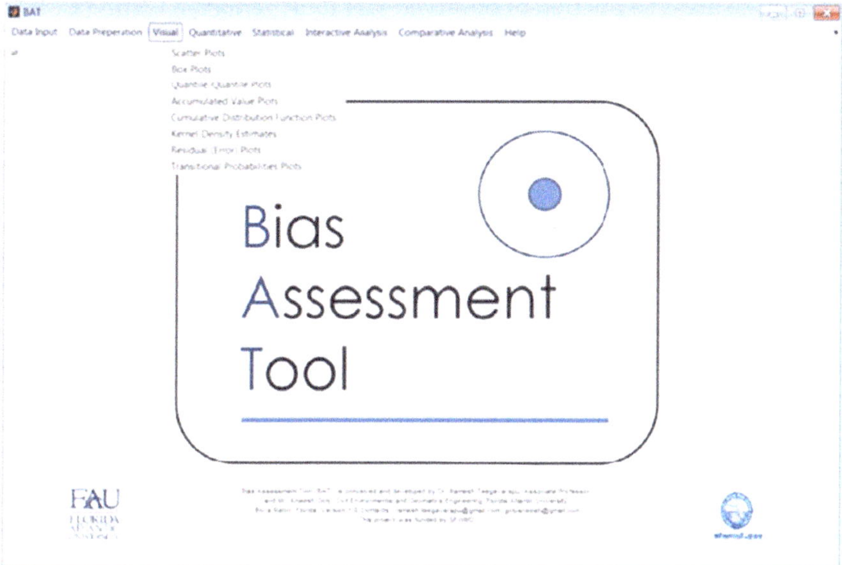

Figure 7-2. Main screen of bias assessment tool (BAT).
Source: Teegavarapu (2012a).

cated by rain-gauge observation being correct based on radar-based estimation to the odds of a wet event estimated by radar data being wrong? The score measures the ratio of the odds of making a hit to the odds of making a false alarm.

An example of bias assessment tool developed by Teegavarapu (2012b) is shown in Figure 7-2.

Contingency measures such as concordance, error rate, specificity, and sensitivity can also be used. The concordance index, which is also referred to as the PC, gives the fraction of all wet and dry time periods correctly estimated based on radar-based rainfall data given observed rain-gauge data. The error rate gives the fraction of all wet and dry time periods incorrectly estimated based on radar-based rainfall data given observed rain-gauge data. Specificity provides information about the fraction of dry time periods obtained by radar-based rainfall estimates that were actually observed as dry time periods based on rain-gauge observations. Sensitivity is also referred to as the success rate. It provides information about what fraction of the wet time periods obtained by radar-based rainfall estimates that were actually observed based on rain-gauge observations.

The bias score gives the ratio of the frequency of wet time periods obtained from radar-based rainfall estimates to the frequency of wet time periods obtained from rain-gauge observations. The POD, also known as the hit rate, measures the fraction of observed wet time periods recorded by rain gauge that was correctly estimated by radar-based estimates. The probability of false detection (POFD), also known as the FAR, measures the fraction of observed dry time periods by rain gauge to the dry events that were defined as wet time periods by radar-based estimates. This measure also gives the fraction of wet time periods based on radar-based rainfall that were observed to be dry time periods.

To develop confidence intervals for performance measures or indexes, bootstrap samples (Wilks 2011) can be generated. These samples are observed rain-gauge and estimated radar-based rainfall data pairs that are drawn randomly with replacement from the available data sets. The performance measures or indexes are calculated for each bootstrap sample for evaluation of radar-based rainfall data. The methodology is shown in Figure 7-3. Confidence intervals (CIs) can be developed for the error performance measures. The intervals can be developed using one of the four methods: (1) basic percentile, (2) bias corrected, (3) bias corrected and accelerated, and (4) studentized. The development of CIs will help in assessing the differences in model performance statistically.

All the error measures can be evaluated by combining them into one score. However, having one combined score may not provide details of individual measures, and there is difficulty in deciding weighting factors for each measure in such a combined score. Combining several measures into one single graph such as Taylor's diagram (Taylor 2001) can be explored. This diagram can help to evaluate the RMSE, the CC, and the standard deviations based on data sets simultaneously. An example of a Taylor diagram for comparison of near-real-time (NRT), end of the day (EOD), and end of

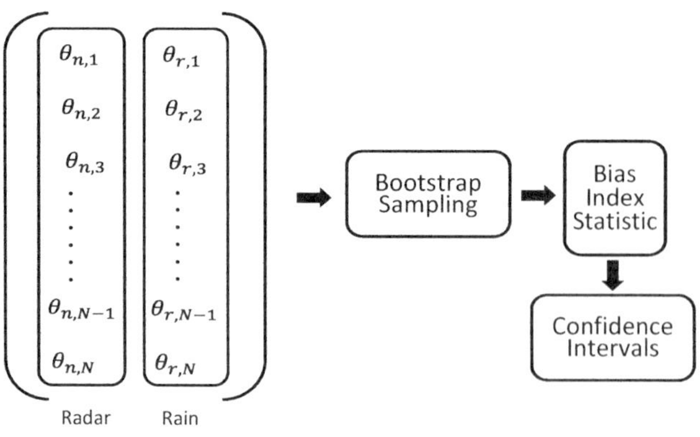

Figure 7-3. Steps for evaluation of confidence intervals of bias statistics using bootstrap sampling.
Source: Teegavarapu et al. (2015).

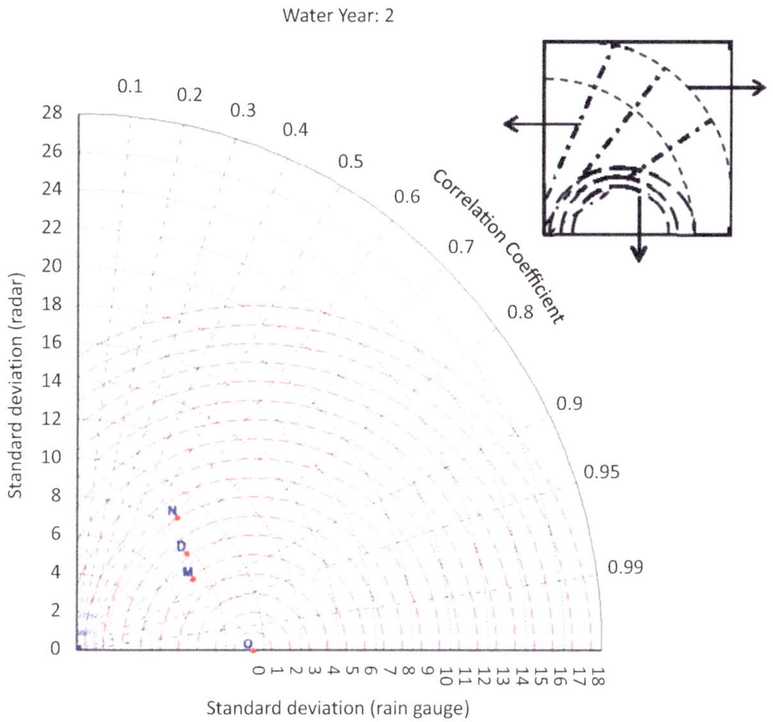

Figure 7-4. Taylor diagram for comparative assessment of NRT (N), EOD (D), EOM (M) radar data and observed (O) rain-gauge observations based on annual totals at 189 sites in South Florida.
Source: Teegavarapu et al. (2015).

the month (EOM) radar data and observed (O) rain-gauge data is provided in Figure 7-4. The radar data is bias corrected over time starting with NRT to the EOM.

Quantitative indexes for evaluation of rain-gauge and radar data will involve calculation of summary statistics such as mean, standard deviation, the coefficient of variation, skewness, kurtosis, trimmed means, and others. Comparative analysis of distributions of radar and rain-gauge data can be carried out using nonparametric statistical hypothesis tests, and they include the (1) Kolmogorov–Smirnov (KS) two-sample test, the (2) Ansari–Bradley test, and the (3) Wilcoxon Rank Sum test. The KS test can be used for testing differences in mean or variance or both. The test is ideal for evaluating overall differences in the two samples being evaluated. The Ansari–Bradley test is ideal for testing differences in variance, with the assumption that the two samples have identical means. The Wilcoxon Rank Sum test evaluates the differences in medians, with the assumption of identical variances.

Quantile–quantile plot (Q–Q plot) can be used as a graphical procedure to compare two distributions based on rain and radar data sets. A Q–Q plot is essentially a plot of the quantiles of two distributions against each other. Comparison of cumulative plots of radar-based rainfall and rain-gauge observations can provide an assessment of time-consistent accumulations of these two data sets. The similarity of two probability density functions (PDFs) can be established using metrics that compare the distances between these two functions. Distance measures used for comparison of the PDFs of radar and rain-gauge data can be used. PDFs can be developed using parametric and nonparametric models using one of the standard distributions or kernel density estimates. However, histograms using constant bins (i.e., class intervals) can be used for comparison (refer to Figure 7-5). The ideal value of the distance between two histograms for exact match or similarity is zero. The different spatiotemporal sampling properties of radar compared with gauges means that even if both instruments have zero measurement error, the distance between the two histograms could vary. The distance metrics calculated over different temporal horizons (e.g., water year) can provide an assessment of similarities in histograms of radar and rain-gauge data. Evaluation of distances will provide an assessment of improvement or worsening of radar data over time. The similarity of probability distributions of rain-gauge and radar-based rainfall data may be required in hydrologic modeling tasks.

The variables P_1, P_2, . . . , P_{nb} and Q_1, Q_2, ... , Q_{nb} refer to relative frequency values obtained using rain-gauge and radar data sets for a prespecified constant value of number of bins. Transition probabilities referred to as two-state first-order Markov chain probabilities can be calculated to evaluate dry and wet spells. The probabilities refer to the probability of the occurrence of positive rainfall and no rainfall in the current time interval given the occurrence of positive rainfall in the previous interval. Similarly,

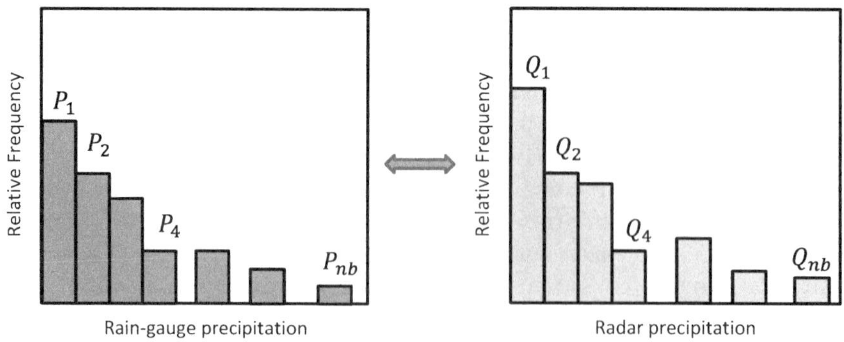

Figure 7-5. Schematic showing two histograms for comparison using distance metrics.

the probabilities refer to the probability of no rainfall in the current time interval given the occurrence of positive rainfall and no rainfall in the previous interval. A comparison of these probabilities based on rain and radar data should reveal the discrepancies in the persistence of wet and dry spells. Probabilities obtained by good-quality radar-based data and rain-gauge observations should be similar. An example of transition probability values for rain and radar data for four water years based on a study in South Florida is shown in Figure 7-6.

Autocorrelation plots of rain and radar data sets can help assess serial correlation, which refers to the correlation between a given variable and itself over several time periods. Sample autocorrelation plots for radar and rain-gauge data for several lagged time intervals are shown in Figure 7-7. It is evident from the figure that the radar data overestimates the serial autocorrelation at several lags relative to gauges. The overestimation by radar may be because the radar is sampling a much large volume of space compared

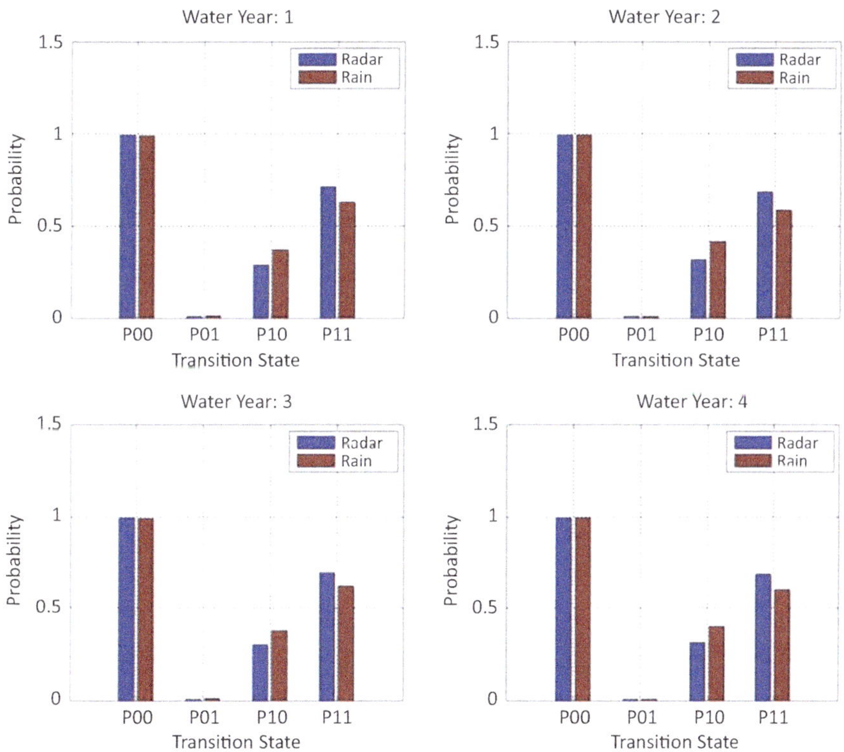

Figure 7-6. Average transition probabilities based on rain-gauge data and radar data at all sites for different water years.
Source: Teegavarapu et al. (2015).

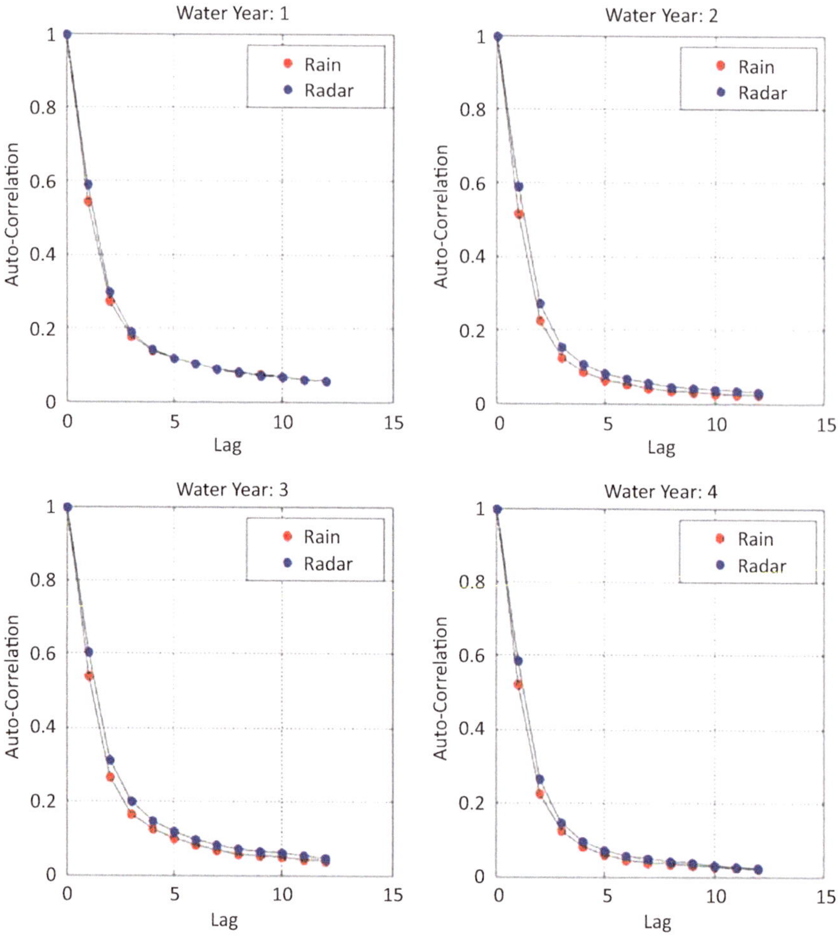

Figure 7-7. Autocorrelation values for radar and rain-gauge data at different lags for four different water years.

to the small spatial scale over which rain-gauge measurements are carried out.

7.2 IDEAL PERFORMANCE MEASURES AND SKILL SCORES

The possible range of values for indexes and skill scores along with ideal or perfect values (Teegavarapu et al. 2015) are provided in Table 7-2. The ideal values will help in the assessment of bias-corrected radar values. Table 7-3 provides the summary of all the indexes, measures, and scores and their utilities for bias analyses.

Table 7-2. Range and Perfect Score Values for Different Indexes or Measures

Index/Measure or Skill Score	Possible Range of Values	Perfect (Ideal) Score
Mean Error (ME)	$-\infty$ to $+\infty$	0
Mean Absolute Error (MAE)	0 to $+\infty$	0
Mean Squared Error (MSE)	0 to $+\infty$	0
Root Mean Squared Error (RMSE)	0 to $+\infty$	0
Nash-Sutcliffe Efficiency Coefficient (NSEC)	0 to 1	1
Correlation Coefficient (ρ [Pearson]	−1 to 1	1
Spearman Correlation Coefficient (ρ_{rank})	−1 to 1	1
Concordance	0 to 1	1
Error Rate	0 to 1	0
Specificity	0 to 1	1
Sensitivity	0 to 1	1
Probability of Detection (POD)	0 to 1	1
Probability of False Detection (POFD)	0 to 1	0
Bias Score (BS)	0 to $+\infty$	1
Heidke Skill Score (HSS)	$-\infty$ to 1	1
Critical Success Index (CSI)	0 to 1	1
Peirce's Skill Score (PSS)	−1 to 1	1
Gilbert's Skill Score (GSS)	−1/3 to 1	1
Odd's Ratio	0 to $+\infty$	$+\infty$

7.3 UTILITY OF ASSESSMENT INDEXES AND PERFORMANCE MEASURES

This section briefly discusses the utility of different measures and indexes for real-time operations, modeling, and short-term planning. The utility of the indexes, measures, and tests are described in Table 7-4. Although a specific index or performance measure is attached to one of the categories, the end user can use the indexes in any way appropriate for decision making.

Table 7-3. Measures and Their Utility

Measures, Indexes, and Evaluation Tests	Utility
Contingency Measures	Assessment of estimates using binary evaluations (states of the system)
Skill Scores	Assess the skill of method/estimate of radar rainfall
Performance Measures (Error Measures)	Measures for evaluation of overall radar estimation accuracy and capabilities
Transitional Probabilities	Evaluation of persistence for different temporal state transitions
Association Measures and Plots	Bi-variate assessment of data
Taylor Diagram	Assessment of multiple radar estimates and performances of models over time
Summary Statistical Measures	Simple measures for testing similarity in central tendency, range, and variability between two data sets
Data Distribution Analysis	Similarity of distributions of radar and rain gauge data
Serial Autocorrelation	Persistence at different temporal intervals
Quantile Plots	Similarity of two data distributions (rain and radar data)
Variograms	Assess spatial variability of data
Bootstrap Sampling-Based Confidence Intervals	Assessment of errors based on re-sampling of error measures
Infilling or Replacement Tests	Validity of data for replacing missing data
Evaluation of Residuals	Check for method bias, structure (temporal), and assessment of normality
Cumulative Plots	Check for consistency/deviations in data sets over time
Kernel Density Estimates	Non-parametric assessment of data for comparisons
Statistical Hypothesis Tests	Two sample tests for confirmation of distribution characteristics of data

Source: Teegavarapu et al. (2015).

Table 7-4. Measures and Their Utility for Operations, Modeling, and Short-Term Planning

Measures, Indexes, and Evaluation Methods	Utility for Operations [O], Modeling [M], and Short-Term Planning [S]
Contingency Measures	Operations/short-term decision making [O, S]
Skill Scores	Short-term planning [S]
Performance Measures (Error) Measures	Modeling, long-term assessment [M]
Transitional Probabilities	Short-term evaluation and modeling [S, M]
Association Measures Plots	Operations, short-term evaluation, modeling [O, S, M]
Taylor Diagram	Modeling and long-term assessment [M]
Summary Statistical Measures	Short-term planning and improvement and modeling [S, M]
Data Distribution Analysis	Modeling [M]
Serial Autocorrelation	Modeling and long-term evaluation [M]
Quantile Plots	Modeling [M]
Spatial Variograms	Modeling [M]
Bootstrap Sampling-Based Confidence Intervals	Modeling [M]
Infilling or Replacement Tests	Operations, short-term planning [O, S]
Evaluation of Residuals	Long-term assessment, modeling [M]
Cumulative Plots	Modeling [M]
Kernel Density Estimates	Modeling [M]
Statistical Hypothesis Tests	Short-term evaluation, modeling [S, M]

Source: Teegavarapu et al. (2015).

It is important to recognize several critical issues and limitations associated with the evaluation of radar-based rainfall estimates using several qualitative and quantitative indexes briefly discussed earlier. Many indexes described in this chapter are based on measures used in forecast evaluation procedures common in the fields of atmospheric sciences and numerical weather prediction and forecast verification. In the arena of forecast verification, forecasts are considered independent of observations. Radar-based rainfall estimates can be considered independent of rain-gauge observations if bias correction is not carried out. In general, real-time or non-real-time radar data are not completely independent from rain-gauge observations.

The following issues are important to consider when carrying out bias analyses using some of the evaluation metrics discussed in this chapter:

- Radar-based rainfall estimates are considered as "model estimates" or "forecasts" as these estimates are surrogate measurements of observed rainfall amounts. This assumption also helps in the application of forecast verification indexes for evaluation of radar-based data.
- Radar-based rainfall estimates are adjusted based on rain-gauge observations. However, radar-based rainfall estimates are not adjusted for each pixel (or spatial grid) in which the rain gauge is located, and the adjustments are not based on one single rain gauge. Bias correction procedures are generally applied to radar-estimated data based on a specific Z–R (reflectivity–rainfall rate) relationship and by using rain-gauge data from a spatial network of gauges.
- Observed rainfall data available from rain gauges are not completely error-free. However, the data available from the rain gauges after preliminary quality assurance and control are assumed to be ground truth and error-free without any systematic errors. The rain-gauge data should be critically evaluated before it can be used for bias analysis.
- It is expected that radar-based observations improve over time, starting from NRT (e.g., from 5 to 15 min) to EOD or at coarser temporal resolutions (month, year). Products available beyond real time are bias corrected and are available after a specified time interval (e.g., day, month, or season). This is mainly owing to improved rain-gauge data quality available from the time of collection to the point in time when these data are available for adjustment of radar data.
- Radar-based rainfall estimates are surrogate measurements or estimates of observed rainfall amounts. This assumption also helps in the application of forecast verification indexes for the evaluation of radar-based data.
- Two-sample nonparametric statistical tests described in this chapter are valid only if the samples from the two data sets are independent.

Because radar data is adjusted based on rain-gauge data, the assumption of independence is not valid to carry out these tests. Radar data that is not bias corrected or adjusted using rain-gauge data can be used with these tests.
- Rain-gauge observations that are not used for adjustment of radar data are ideal for analysis of biases of radar data.

Ideally, the indexes or measures described in this chapter are suitable for radar-based rainfall estimates that have not undergone any bias corrections using rain-gauge data.

7.4 BIAS CORRECTIONS

Bias corrections refer to corrections applied to radar-based (or in some cases, satellite-based) rainfall data using the rain gauge. Optimization formulations (Teegavarapu 2012a) can be developed and solved for obtaining optimal parameters in the bias correction methods. The objective in such formulations is to minimize the absolute error based on rain-gauge observations and bias-corrected radar-based values. Local correction refers to the use of a select number of gauges that are part of all the available gauges in a network. Correction that uses all gauges in the region or network is referred to a global correction. Readers are referred to recent works of Wright et al. (2013) and Habib et al. (2013) for exhaustive discussion on bias correction methods and applications.

The parameters optimized are (1) distance, (2) friction factor (or exponent used for the distance, and (3) upper limit value on the gauge to radar (G/R) observations ratio. Several questions need to be answered before bias correction procedures are attempted:

- What are the improvement methods (bias correction) that can be used?
- What are the spatial and temporal resolutions of the rain-gauge data that can be used?
- What is the best bias correction (local or global)?
- If local correction is planned, how many rain gauges should be used?
- If local correction is planned, can the number of rain gauges be optimally selected for a given period (e.g., season, etc.)? The optimal selection would entail selection of well-functioning gauges over the domain of interest.
- If local correction is planned, what is the spatial extent to be considered?
- Are bias correction methods different for extreme events?

7.5 BIAS CORRECTIONS WITH LIMITED RAIN-GAUGE DATA

In general, bias estimation and correction procedures are undertaken using a number of rain-gauge stations. Collocated rain-gauge and radar pixel (grid) data are used to evaluate the improvements after bias corrections are completed. An example of a general way to assess the number of gauges needed for bias correction is discussed by Wright et al. (2013). It is not necessary for all gauges used for bias correction to have useful data for all time periods. Any erroneous gauge data can be purged from the data set before bias correction is carried out. Rain-gauge consistency can be evaluated using double mass/double accumulation curves.

Performance evaluations can be compared by using empirical cumulative density functions of the CC and absolute errors or any other performance measure. In bias correction procedures, one of the main problems encountered with the rain-gauge data is the lack of consistent records of observations. In many instances, the time series data within a time frame (e.g., a month or a year) is not of constant length for all the rain gauges. Lack of constant length of rain-gauge data poses difficulties in implementing the bias correction methods, as consistent record lengths are crucial. Rain-gauge selection should be carried out, and the temporal resolution must be decided before any bias correction methods are applied to the radar data for improvement. Rain-gauge stations should be selected based on two criteria: (1) number of gauges and (2) consistency of rain-gauge data quality.

7.6 BIAS CORRECTIONS: TEMPORAL RESOLUTION ISSUES

In general, bias correction methods require accurate rain-gauge data observations without any gaps at a temporal resolution equivalent to that of radar data. However, data gaps for individual gauges within a network are acceptable, as long as a reasonable number of gauges are available for a given period. Missing data should be filled before the corrections procedures are adopted. Rainfall data from rain gauges should be thoroughly evaluated for any anomalies. Bias corrections may involve the use of inverse distance and exponential weighting methods. Optimal selection of parameters for weighting methods is possible and may not always be computationally efficient or tractable for corrections at time resolutions less than one day. MFB correction procedures may sometimes yield better performance compared to other local weighting methods (Habib et al. 2013). Therefore, the MFB procedure should always be tested along with more conceptually complex methods. A comparative analysis of bias correction methods using different temporal resolutions (e.g., 15 min, hour, and day) is recommended whenever rain-gauge and radar data are available at these resolutions. The absence of rain-gauge data at the same resolution as radar data (e.g., 15 min data) forces bias corrections based on the coarser time scale (e.g., day or hour) to be applied to data at a finer time scale (e.g., 15 min or less). In general, improvement in bias corrections can be realized only if data at similar temporal resolutions for rain gauge and radar are used. However, MFB corrections based on coarser temporal resolutions applied to finer resolution data have also provided improvements (Wright et al. 2013).

7.7 CONCLUSIONS

Assessment of bias in radar-based precipitation estimates compared to rain-gauge observations is a critical exercise to evaluate pre- and post-corrected (gauge-adjusted) radar-based precipitation data. A framework for bias analysis and corrections and issues related to bias assessment are presented in this chapter. The framework is not exhaustive enough to cover the range of potential needs and techniques required for assessment of bias in radar-based precipitation estimates.

REFERENCES

Anagnostou, E. N., W. F. Krajewski, D. Seo, and E. R. Johnson. 1998. "Mean-field rainfall bias studies for WSR-88D." *J. Hydrol. Eng.* 3(3), 149–159.

Bedient, P. B., W. C. Huber, and B. E. Vieux. 2013. *Hydrology and floodplain analysis*. 6th Ed. Upper Saddle River, NJ: Prentice-Hall.

Borga, M., E. N. Anagnostou, and E. Frank. 2000. "On the use of real-time radar rainfall estimates for flood prediction in mountainous basins." *J. Geophys. Res.: Atmos.* 105(D2), 2269–2280.

Borga, M. 2002. "Accuracy of radar rainfall estimates for streamflow simulation." *J. Hydrol.* 267(1-2), 26–39.

Brandes, E. A. 1975. "Optimizing rainfall estimates with the aid of radar." *J. Appl. Meteor.* 14, 1339–1345.

CAWCR (Center for Australian Weather and Climatic Research). 2018. Center for Australian Weather and Climatic Research. Accessed August 2018. http://www.cawcr.gov.au/projects/verification/.

Dinku, T., E. N. Anagnostou, and M. Borga, M. 2002. "Improving radar-based estimation of rainfall over complex terrain." *J. Appl. Meteorol.* 41(12), 1163–1178.

Germann, U., G. Galli, M. Boscacci, and M. Bolliger. 2006. "Radar precipitation measurement in a mountainous region." *Q. J. R. Meteorolog. Soc.* 132(618), 1669–1692.

Grassotti, C., R. N. Hoffman, E. R. Vivoni, and D. Entekhabi. 2003. "Multiple-timescale intercomparison of two radar products and rain gauge observations over the Arkansas–Red River basin." *Weather Forecast.* 18, 1207–1229.

Haberlandt, U. 2007. "Geostatistical interpolation of hourly precipitation from rain gauges and radar for a large-scale extreme rainfall event." *J. Hydrol.* 332, 144–157.

Habib, E., and W. F. Krajewski. 2002. "Uncertainty analysis of the TRMM ground-validation radar-rainfall products: Application to the TEFLUN-B field campaign." *J. Appl. Meteorol.* 41(5), 558–572.

Habib, E., L. Qin, D.-J. Seo, G. J. Ciach, and B. R. Nelson. 2013. "Independent assessment of incremental complexity in NWS multisensor precipitation estimator algorithms." *J. Hydrol. Eng.* 18, 143–155.

Jayakrishnan, R., R. Srinivasan, and J. G. Arnold. 2004. "Comparison of raingage and WSR-88D Stage III precipitation data over the Texas-Gulf basin." *J. Hydrol.* 292(1), 135–152.

Johnson, D., M. Smith, V. Koren, and B. Finnerty. 1999. "Comparing mean areal precipitation estimates from NEXRAD and rain gage networks." *J. Hydrol. Eng.* 4(2), 117–124.

Jolliffe, I. T., and D. B. Stephenson. 2012. *Forecast verification: A practitioner's guide in atmospheric science.* Hoboken, NJ: Wiley-Blackwell.

Kitchen, M., R. Brown, and A. G. Davies. 1994. "Real-time correction of weather radar data for the effects of bright band, range and orographic growth in widespread precipitation." *Q. J. R. Meteorolog. Soc.* 120(519), 1231–1254.

Krajewski, W. F., and J. A. Smith. 2002. "Radar hydrology: Rainfall estimation." *Adv. Water Res.* 25(8-12), 1387–1394.

Livneh, B., et al. 2013. "A long-term hydrologically based dataset of land surface fluxes and states for the conterminous United States: Update and extensions." *J. Clim.* 26(23), 9384–9392.

Lopez, P. 2011. "Direct 4D-Var assimilation of NCEP stage IV radar and gage precipitation data at ECMWF." *Mon. Weather Rev.* 139(7), 2098–2116.

Lott, N., and M. Sittel. 1996. "A comparison of NEXRAD rainfall estimates with recorded amounts." *Tech. Rep. 96-03.* Ashville, NC: National Climatic Data Center.

Maurer, E. P., A. W. Wood, J. C. Adam, D. P. Lettenmaier, and B. Nijssen. 2002. "A long-term hydrologically-based data set of land surface fluxes and states for the conterminous United States." *J. Clim.* 15(22), 3237–3251.

Nash, J. E., and J. V. Sutcliffe. 1970. "River flow forecasting through conceptual models, Part-1: A discussion of principles." *J. Hydrol.* 10, 282–290.

Pathak, C. S., M. Onderlinde, and H. E. Fuelberg. 2009. "Use of NEXRAD rainfall data to develop climatologically homogenous rain areas for central and south Florida." *Proc., EWRI World Environmental and Water Resources Cong.*, Reston, VA: ASCE, 1–11. DOI: 10.1061/41036(342)618.

Seo, D.-J., J. P. Breidenbach, and E. R. Johnson. 1999. "Real-time estimation of mean field bias in radar rainfall data." *J. Hydrol.* 223(3), 131–147.

Seo, D.-J., and J. Breidenbach. 2002. "Real-time correction of spatially non-uniform bias in radar rainfall data using rain gauge measurements." *J. Hydrometeor*. 3, 93–111.

Steiner, M., J. A. Smith, S. J. Burges, C. V. Alonso, and R. W. Darden. 1999. "Effect of bias adjustment and rain gage data quality control on radar rainfall estimation." *Water Resourc. Res.* 35(8), 2487–2503.

Taylor, K. E. 2001. "Summarizing multiple aspects of model performance in a single diagram." *J. Geophys. Res.* 106(D7), 7183–7192.

Teegavarapu, R. S. V. 2012a. "Assessment and evaluation of NEXRAD-based rainfall estimates in South Florida Water Management District." *SFWMD Rep.*, West Palm Beach, FL: SFWMD.

Teegavarapu, R. S. V. 2012b. "Bias Assessment Tool (BAT), Florida Water Management District." *SFWMD Rep.*, West Palm Beach, FL: SFWMD.

Teegavarapu, R. S. V., Goly, A., Wu, Q. 2015. "Comprehensive framework for assessment of radar-based precipitation data estimates." *J. Hydrol. Eng.* 10.1061/(ASCE)HE.1943-5584.0001277, E4015002.

Vignal, B., G. Galli, J. Joss, and U. Germann. 2000. "Three methods to determine profiles of reflectivity from volumetric radar data to correct precipitation estimates." *J. Appl. Meteorol.* 39(10), 1715–1726.

Wilks, D. S. 2011. *Statistical methods in the atmospheric sciences*. Cambridge, MA: Academic Press.

Wright, D. B., J. A. Smith, G. Villarini, and M. L. Baeck. 2013. "Long-term, high-resolution, radar rainfall records for urban hydrology." *J. Am. Water Resour. Assoc.* 50(3), 713–734.

Xiao, Q., and J. Sun. 2007. "Multiple-radar data assimilation and short-range quantitative precipitation forecasting of a squall line observed during IHOP_2002." *Mon. Weather Rev.* 135(10), 3381–3403.

Yilmaz, K. K., T. S. Hogue, K. Hsu, S. Sorooshian, H. V. Gupta, and T. Wagener. 2005. "Intercomparison of rain gage, radar, and satellite-based precipitation estimates with emphasis on hydrologic forecasting." *J. Hydrometeorol.* 6(4), 497–517.

Young, C. B., B. R. Nelson, A. A. Bradley, J. A. Smith, C. D. Peters-Lidard, A. Kruger, et al. 1999. "An evaluation of NEXRAD precipitation estimates in complex terrain." *J. Geophys. Res.* 104(D16), 19691–19703.

Young, C., A. A. Bradley, W. F. Krajewski, A. Kruger, and M. Morrissey. 2000. "Evaluating NEXRAD multisensor precipitation estimates for operational hydrologic forecasting." *J. Hydrometeorol.* 1, 241–254.

Young, C. B., and N. A. Brunsell. 2008. "Evaluating NEXRAD estimates for Missouri River basin: Analysis using daily rain gage data." *J. Hydrol. Eng.* 7, 549–553.

CHAPTER 8

ADVANCED TOPIC: RAIN-GAUGE RAINFALL DATA AUGMENTATION AND RADAR RAINFALL DATA ANALYSIS

8.1 SPATIAL AND TEMPORAL ANALYSIS OF RAINFALL

Continuous rainfall data without any gaps are needed for hydrologic modeling. Often rainfall data gaps of different lengths are unavoidable because of random or systematic errors. Incorrect recording and transcription of rainfall data creates gaps in the data that need to be filled and casts doubt on the reliability of the data for statistical and trend analysis (Hosking and Wallis 1998, Wallis et al. 1991, Teegavarapu et al. 2011, Teegavarapu and Nayak 2017). Temporal and spatial interpolation methods are often used in infilling or imputation of missing rainfall data. Temporal interpolation using any autoregressive technique is possible only when serial autocorrelation is high enough for several temporal lags. In general, infilling of missing rainfall data for temporal resolutions with low or negligible serial autocorrelation at different lags requires spatial interpolation. Spatial interpolation using local or global variants, exact or inexact versions, and stochastic and deterministic techniques are used for estimation of missing rainfall data (Teegavarapu and Chandramouli 2005, Teegavarapu 2012b). Local and global variants refer to interpolation methods that use a subset of all observation sites and all sites, respectively. Exact interpolators predict the value at a site that is the same as the observed value, but inexact interpolators cannot. Deterministic interpolation techniques cannot provide estimation error, whereas stochastic methods can. The missing rainfall data are considered missing at completely random (MCAR), and estimation techniques use observed data at different control points in space to fill values at a given site. The following section provides a brief overview of missing data estimation methods for rainfall records.

Deterministic weighting and stochastic interpolation methods have been used for the spatial construction of rainfall fields or for estimating missing rainfall values in space in the past in a number of studies. Weighting methods belong to a class of spatial interpolation techniques such as inverse distance (Simanton and Osborn 1980, Wei and McGuinness 1973), nonlinear deterministic and stochastic interpolation methods (e.g., kriging). Regression and time series analysis methods are examples of data-driven approaches. Global interpolation methods that use trend surface analysis and regression provide several advantages compared to deterministic weighting techniques. Trend surface analysis uses a polynomial equation of spatial coordinates to approximate points with known values. These methods and their several variants are discussed in the next few sections.

The inverse distance weighting method (IDWM) is the most commonly used approach for estimating missing data and is often referred to as the NWS method (ASCE 1996). In the field of quantitative geography, IDWM is used for spatial interpolation (O'Sullivan and Unwin 2010). Hodgson (1989) modified IDWM to include a learned search approach that reduces the number of distance calculations. To incorporate topographical aspects, Shepard (1968) proposed a modified IDWM that is referred to as a barrier method. Variance-dependent surface interpolation methods, belonging to the general family of kriging, have been applied to hydrological interpolation problems (Vieux 2001, Grayson and Bloschl 2001). These interpolation schemes are based on the principle of minimizing the estimate of variance at points where measurements are unavailable. Kriging in various forms has been used to estimate missing rainfall data at stations, as well as to interpolate rainfall from point measurements (Dingman 2007, Vieux 2001). Ashraf et al. (1997) compared interpolation methods (kriging, inverse distance, and co-kriging) to estimate missing rainfall values. They indicated that kriging provided the smallest RMSE. Co-kriging of radar and rain-gauge data was used by Krajewski (1987) to estimate mean areal rainfall. Seo et al. (1990a, 1990b) described the use of co-kriging and indicator kriging for interpolating rainfall data. The reader is referred to Teegavarapu (2016) for a comprehensive review of different spatial interpolation methods and their limitations.

8.2 MISSING DATA ESTIMATION

Teegavarapu and Chandramouli (2005) and Teegavarapu (2012b) reported several limitations and advantages of using deterministic and stochastic spatial interpolation techniques to estimate missing rainfall data at a base gauge (i.e., a station with missing data) using data at all other stations. They indicated that all interpolation techniques would provide inaccurate estimates of missing rainfall data in two situations: (1) when rainfall is measured at one or more stations but no rainfall actually occurred at the base gauge and (2) when rainfall occurs at the base station but no rainfall is measured or occurred at all other stations. In Case 1, all spatial interpolation techniques will produce a positive estimate, although in reality, zero rainfall is recorded at the base station. It is impossible to estimate missing rainfall data in the second case because point observations are used to estimate the missing value at the base station using spatial interpolation algorithms alone. All interpolation techniques produce a zero estimate for the situations encountered in Case 2. Data from other sources (e.g., radar-based estimates) can be used in these situations to estimate the missing values.

Limitations of spatial interpolation methods have been reported in recent studies. Vieux (2001) pointed out several limitations of the IDWM, with a major one being the "tent pole effect" that leads to greater estimates closer to the point of interest. Grayson and Bloschl (2001) list several limitations of Thiessen polygons and inverse distance methods. They suggested that these methods should not be used for spatial interpolation considering their limitations related to the inability of polygon-based methods to generate smooth surfaces and invalid use of Euclidean distance as a surrogate for the strength of correlation between two observation sites. However, they recommend thin-splines and kriging for interpolating hydrologic variables. The Thiessen polygon approach has the major limitation of not providing a continuous field of estimates when used for spatial interpolation (O'Sullivan and Unwin 2010, Teegavarapu 2016). Brimicombe (2003) indicated that the main point of contention in applying IDWM to spatial interpolation is selecting the number of relevant observation points used for the spatial interpolation. Correlation weighting techniques and artificial neural network (ANN) methods are proved conceptually to be superior to deterministic approaches compared to traditional IDWM and its variants (Teegavarapu and Chandramouli 2005). The kriging estimation method (KEM) is considered a reliable interpolation technique (O'Sullivan and Unwin 2010), but it has several limitations. These include selecting the appropriate semi-variogram, assigning objective values to the sill and nugget parameters, and distance intervals at which variance estimates are obtained. Distance weighting methods that are often used for estimating missing rainfall records were revisited by Teegavarapu and Chandramouli (2005). Conceptual revisions of the methods addressed two main issues relating to the definition of distance used in the calculations and the process of selecting the nearby gauges. Universal functional approximators such as ANNs were used for fitting a semi-variogram model using the raw data in ordinary kriging to estimate missing rainfall data by Teegavarapu (2007a). The use of ANN eliminates the need for predefined semi-variogram models to capture the spatial variation of data and the trial and error process involved in estimation of semi-variogram parameters. The association rule mining (ARM)-based spatial interpolation approach was recently discussed by

Teegavarapu (2007b) to improve the rainfall estimates provided by deterministic and stochastic spatial interpolation techniques. Considerable improvements in the estimates were achieved when ARM is used in conjunction with other interpolation techniques. Teegavarapu et al. (2009) developed optimal functional forms for estimation of missing precipitation data.

Range- and cluster-based optimization methods in space and time were developed by Teegavarapu and Bajaj (2008) and Teegavarapu (2012a, 2014). Optimization models using mixed-integer linear and nonlinear programming (MINLP) formulations have been developed by Teegavarapu (2012a) for estimation of missing rainfall data at a gauge. These formulations use binary variables for the selection of rain gauges that participate in the spatial interpolation process and also in the selection of an optimum cluster of rain gauges for estimation of missing data. Several variants of these mathematical programming formulations have been proposed by Teegavarapu et al. (2005) and Teegavarapu (2012a) to improve the estimates of missing rainfall data. The variants investigated in their studies [Teegavarapu et al. (2005), Teegavarapu (2012a)] improved estimates of rainfall data compared to those obtained from traditional distance-based weighting methods. A nonlinear mathematical programming model using binary variables was proposed and investigated by Teegavarapu and Pathak (2008) to infill missing rainfall records using radar (NEXRAD)-based rainfall estimates. The model identifies the cluster of radar data values that can be used for infilling the rain-gauge records.

8.3 USE OF RADAR DATA FOR INFILLING RAINFALL DATA

8.3.1 Functional Forms Linking Radar and Rain-Gauge Data

When available, radar-estimated precipitation data can be used for infilling missing rain-gauge data. Several issues that affect the infilling methods include the availability of historical rain-gauge and radar data, the spatial and temporal variability of rainfall, radar–rain gauge relationships, and selection of spatial extent of radar data. Before embarking on rain-gauge data filling using radar, one should evaluate the influence of the spatial and temporal variability of rainfall processes on the performance of spatial interpolation algorithms. Homogeneous rainfall areas defined based on rainfall processes can help in the development of spatially varying (i.e., local) methods for infilling of missing rain-gauge data.

Seasonal variation of rainfall, rainfall areas that are delineated based on physical processes affecting the genesis and morphology of rainfall processes, and other factors may affect the performance of infilling methods. There is a need to understand and quantify the relationship between the rain gauge and rainfall estimates based on other forms of measurements (e.g., radar) and rainfall measurements at other observation points. Quantification of this relationship is essential to checking the quality of rainfall data from radar and rain gauges. This task is possible by evaluating the radar and rain-gauge measurements independently and then against each other. The importance of such a check was confirmed by a state-of-the-art review of radar rainfall data usage in urban drainage (Einfalt et al. 2004). Similarly, functional relationships can be developed relating rainfall measurements at two or more observation stations. The development of two types of functional forms was discussed by Teegavarapu et al. (2008). These include (1) fixed functional form with predefined functional expressions and (2) flexible functional form–binary formulations. Function-free forms (i.e., universal function approximators) were also tested by Teegavarapu et al. (2008) for linking radar and rainfall observations. It is important to note these formulations use radar as a predictand and rain gauge as a predictor.

Mathematical programming models using linear and nonlinear formulations have been developed. The formulations are aimed to obtain optimal exponents of a power law model and also to obtain optimal functional forms using binary variables. The models, along with constraints adopted from Teegavarapu et al. (2008), are discussed next.

Model I.

$$\text{Minimize} \quad \sqrt{\frac{\sum_{i=1}^{no} (\hat{\phi}_i^m - \phi_i^m)^2}{no}} \tag{8-1}$$

$$\hat{\phi}_i^m = f(\theta_i^m, a, b) \quad \forall i \tag{8-2}$$

$$k_l \le a \le k_u \tag{8-3}$$

$$c_l \le b \le c_u \tag{8-4}$$

where
a and b = exponents
k and c = variable upper and lower bounds on the coefficients, respectively,
$\hat{\phi}_i^m$ = estimate of radar data values,
ϕ_i^m = radar data value for the grid with a specific spatial resolution,
θ_i^m = rainfall value observed on a given day, and
no = number of observations used for obtaining optimal coefficients (i.e., a and b).

Model II. Formulation for model II is exactly similar to Model I with a change in the objective function used for optimization.

$$\text{Minimize} \quad \frac{\sum_{i=1}^{no} |\hat{\phi}_i^m - \phi_i^m|}{no} \tag{8-5}$$

Model III. The formulation for Model III uses the same objective function as that used in Model I with modifications to constraints and the nature of function selection. The selection is possible by use of binary variables.

$$\text{Minimize} \quad \sqrt{\frac{\sum_{i=1}^{no} (\hat{\phi}_i^m - \phi_i^m)^2}{no}} \tag{8-6}$$

$$\hat{\phi}_i^m = f_1(\theta_i^m, a_1, b_1) y_1 + f_2(\theta_i^m, a_2, b_2) y_2 \ldots\ldots + f_k(\theta_i^m, a_k, b_k) y_k \quad \forall i$$

$$\sum_{j=1}^{k} y_k = 1 \tag{8-7}$$

$$kl_i^k \le a_k \le ku_i^k \quad \forall i \tag{8-8}$$

$$cl_i^k \le b_k \le cu_i^k \quad \forall i \tag{8-9}$$

where
a_k and b_k = exponents or constants,
k and c = variable upper and lower bounds on the coefficients or constants, respectively,
$\hat{\phi}_i^m$ = estimate of radar data values,
ϕ_i^m = radar rainfall value for the grid (2 km × 2 km),
θ_i^m i = rainfall value observed on a given day,
no = number of observations used for obtaining optimal coefficients, and
$y_1 \ldots . y_k$ = binary variables.

8.3.2 Artificial Neural Networks

The application of ANNs as universal function approximators has generated enormous interest in the hydrology and water resources research communities (ASCE 2001a, 2001b) in the past several decades. ANNs are

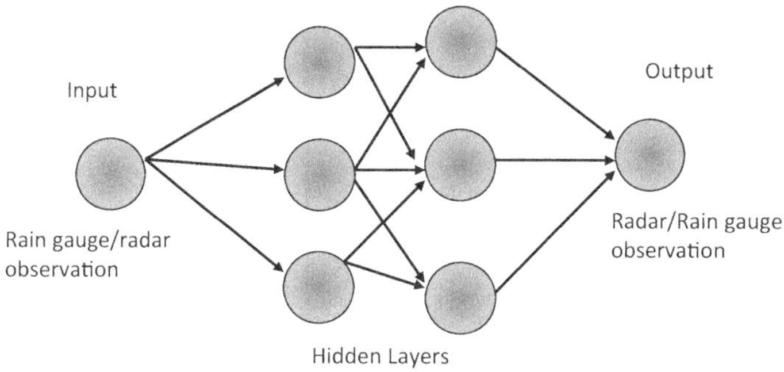

Figure 8-1. Architecture of ANN model used for functional relationship development.
Source: Modified after Teegavarapu et al. (2008).

data-driven approaches that rely on learning relationships between dependent and independent variables to predict the variables of interest. An ANN with two hidden layers and an appropriate number of neurons can approximate any continuous function. This ability has been confirmed using mathematical proofs that were provided independently by Cybenko (1989) and Hornik et al. (1989). Therefore, ANNs are referred to as universal function approximators. French et al. (1992) used a feed forward neural network with a back-propagation training algorithm to forecast fields of rainfall intensity at a lead time of 1 h using the current rainfall field as input. Navone and Ceccatto (1994) used an ANN model to predict summer monsoon rainfall over India. A successful application of using ANN for estimating missing rainfall data recently was reported by Teegavarapu and Chandramouli (2005). In the context of developing radar–rain gauge relationships, ANN models are used to develop functional relationships among rain-gauge and radar data sets. The architecture shown in Figure 8-1 is a two-hidden-layer network that can be used to relate rain-gauge observations and radar-based rainfall estimates. The network can be trained using a back-propagation training algorithm (French et al. 1992).

8.3.3 Geographically Weighted Optimization

Geographically weighted optimization (GWO) (Teegavarapu et al. 2010) is based on the concept of geographically weighted regression (Fotheringham et al. 2002). In GWO, a moving window (fixed number of pixels or grids) is adopted for obtaining optimal weights in a scheme to interpolate missing values at a gauge centered in the pixel in the center of the window. The radar-based rainfall estimates that are available in a grid format surrounding the grid (cell) in which the rain gauge is located are used in the model. The model will identify the cluster of radar data values that can be

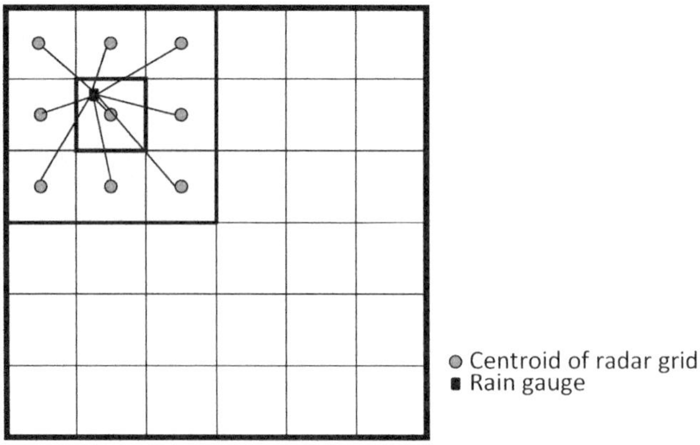

Figure 8-2. Use of 3×3 radar grid for GWO-based estimation of missing data at rain gauge.

used for infilling the rain-gauge records. Spatial and temporal variability of weights for different clusters need to be investigated. The weights can be obtained using a linear or nonlinear optimization formulation. If utilized, Euclidean distances in GWO are based on distances calculated between the centroids of the radar grids and the rain gauges in the optimization schemes (refer to Figure 8-2).

Teegavarapu et al. (2010) documented several variants of weighted interpolation schemes to estimate missing values at rain gauges. The variants are (1) a simple inverse distance–based scheme with optimized exponents, (2) a correlation-based weighting scheme using historical correlation between the rain-gauge and radar observations, (3) an optimal weighting scheme that uses radar data from surrounding grids and rain-gauge data from other rain gauges in the region, and (4) interpolation schemes in which an optimum number of rain gauges and grid-based observations are selected. All these variants were based on a 3×3 grid (a total of nine grids or cells) of radar data with the rain gauge located in the center cell.

8.4 GEOSPATIAL GRID-BASED TRANSFORMATIONS OF RADAR-BASED RAINFALL DATA

Geospatial interpolation methods are often necessary in instances where the rainfall estimates available from multisensor source data on a specific spatial grid need to be transformed to another grid with a different spatial grid or orientation. Spatial interpolation using point rainfall data to generate rainfall fields or surfaces is an essential task in distributed hydrologic modeling. Also, the generation of rainfall data sets confined to a fixed tessellation (grid of specific size) is generally the final product of any processed radar data. Radar data available as gridded data can be used for estimation of missing rainfall data at a rain gauge using local filters (Lloyd 2007) with appropriate focal functions to derive the value of a cell using values from the nearest group of cells. Moving window approaches using focal operators (Lloyd 2007, 2010) are common in image processing problems. The moving window approach was adopted in a recent study (Teegavarapu and Pathak 2008) in which pixels (grids) with radar-based rainfall estimates surrounding a rain gauge were used to estimate missing data at that gauge. Spatial interpolation involving grid operators such as local, focal, zonal, and global functions (Chou 1997) is generally used to process grids for spatial segmentation and classification.

Spatial domain filters (Mather 2004) are used in remotely sensed image processing studies that use moving-average windows to reduce the variability of images. These spatial filters adopt different forms of focal operators (Smith et al. 2007) where the value of any given cell (grid or pixel) is a function of values from the surrounding pixels. However, no optimization formulations are generally incorporated into these filters. Spatial interpolation methods for grid-based transformations of rainfall data are possible by extending the principles of geospatial analysis. An example of the transformation of rainfall estimation from one grid [HRAP (Hydrologic Research Analysis Project) 4 km × 4 km (2.49 mi × 2.49 mi) diagonal] to another [Cartesian 2 km × 2 km (1.24 mi × 1.24 mi) grid] was presented in a study by Teegavarapu et al. (2012) (Figure 8-3). Details of the HRAP grid used for precipitation estimation from WSR-88D radars are provided by NOAA (2013).

The spatial weighting methods used for this example can be referred to as geographically moving window interpolation methods. Studies involving application of grid-to-point interpolation methods were reported recently for establishing rain gauge–radar relationships (Teegavarapu et al. 2008) and infilling of missing rainfall data at a gauge using radar data (Teegavarapu and Pathak 2008). Resampling methods are ideal for problems involving geometric transformation of digital images (Chang 2006, Mather 2004, Lloyd 2010). Three resampling methods—nearest neighbor, bilinear, and cubic convolution—use information from surrounding cells neighboring a point in weighted or nonweighted schemes for interpolation. The nearest-neighbor scheme uses one control point closest to the point where the interpolated value is required and bilinear interpolation uses four nearest neighbors. Two methods used in this example are similar to nearest-neighbor and bilinear interpolation schemes. The similarity is in the use of one or four nearest neighbors and not in the weighting scheme.

The area weighting method uses the areal extent of the overlay as a weight. This method is similar to the natural neighbor method or the area-stealing interpolation (Sibson 1981). In the natural neighbor method, the neighbors of any point are those associated with neighboring Voronoi (Thies-

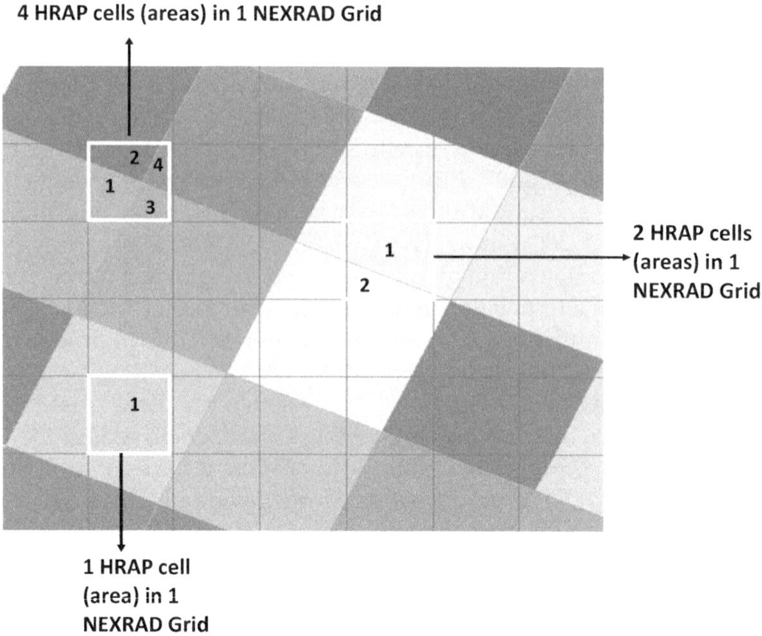

Figure 8-3. Overlay of HRAP and NEXRAD grids.
Source: Teegavarapu et al. (2012).

sen) polygons. Thiessen polygons constructed with and without the point of interest using all the observation points in space provide the proportions of areal overlaps. These proportions are used as the weights. The area weighting method used does not require the construction of Voronoi polygons.

In the maximum area method, the maximum areal extent of the overlay is identified, and the value associated with the maximum area is assigned to the estimate required for the Cartesian grid. This method is similar to the nearest-neighbor method (Mather 2004). However, the nearest neighbors are not selected based on the Euclidean distance but rather on the maximum areal extent of the HRAP grid. This method can be referred to as single best estimator (SBE), as only information from one HRAP grid is always used. The inverse distance weighting is based on the distance from the centroids of the HRAP grids to the centroids of the Cartesian grids. Inverse distance weighting can be constrained to a specific number of neighbors (in this case, four) and can be constrained by a neighborhood region, say, a closed polygon (in this case, a circle of a predefined radius). The method is similar to bilinear interpolation in some aspects. However, the bilinear method uses four nearest neighbors and requires three linear interpolations (Chang 2006). In the case of the IDWM, the number of neighbors need not be fixed, and the reciprocal distances are used as weights. The bilinear interpolation requires more computational time compared to the inverse distance method.

In the equal weights (Average) method, equal weights are assigned to all the HRAP cells that intersect the NEXRAD cells. This is the simplest of all the methods. This method is similar to a moving-average filter used in image processing and is also classified as a spatial domain filter (Mather 2004). In the case of a spatial domain filter, the cells surrounding the central cell are used to obtain a weighted value for the central cell. In the equal weights method, only those HRAP cells that intersect with the central NEXRAD cell grid are adopted for estimation. The methods discussed for grid-based interpolation are radius-limited (i.e., fixed distance), as well as nearest-neighbor local interpolators. Spatial analysis procedures are used to establish the nearest neighbors. The limits on Euclidean distances need not be chosen arbitrarily but through several multiple executions of interpolation methods using different search neighborhoods. Variants of the methods developed will produce different reconstructions of rainfall fields. These variants need to be thoroughly evaluated, as there is no universally best interpolation scheme (O'Sullivan and Unwin 2010). Stochastic interpolation methods based on the concepts of geostatistics can be used for the transformation. Ordinary kriging is one such geostatistical interpolation method that requires the development of semi-variograms for each grid based on the nearest HRAP grid data. Although the development of semi-variograms is not difficult, the selection of an appropriate semi-variogram suitable for the spatial data structure from a set of selected semi-variograms (i.e., exponential, Gaussian, etc.) is time- and computationally intensive, especially when data needs to be transformed for several thousand cells.

8.5 ISSUES WITH FILLED PRECIPITATION DATA SERIES

Infilling methods may introduce significant biases when the gaps constitute more than 20% of the data. A study by Teegavarapu et al. (2011) indicated that bias introduced by infilling methods could alter the statistics of the data. The extent (i.e., temporal length) of filling will affect the results of trend analysis of precipitation extremes (Teegavarapu and Nayak 2017). Spatial interpolation becomes unavoidable when temporal autocorrelations at several lags are negligible, precluding the use of using time series models for infilling. In many instances, gauges that are selected as SBEs based on Euclidean distance or other statistical distances should be used for infilling extreme values. Schuenemeyer and Drew (2010) indicate that any statistical inference made from data with more than 15% of missing observations should be interpreted with caution. Estimation of missing extreme rainfall data is difficult and should not be carried out by global interpolation methods. In many instances, gauges that are selected as SBEs based on Euclidean distance or other statistical distances should be used for infilling extreme values. Infilling of rainfall data at a site should focus on preservation of site-specific and regional statistics. Regional statistics are mainly defined as site-to-site data relationships, spatial autocorrelations (correlations among observations at different sites), and spatial variability of data across the region.

Infilling of missing data in a long rainfall data series introduces the following complexities (Teegavarapu 2013) when conducting frequency analysis:

1. Local and global spatial interpolation methods tend to underestimate higher magnitude events and overestimate lower magnitude events.
2. Site-specific statistics and regional (site-to-site) relationships can be modified.
3. Transition probabilities that characterize persistence and wet-to-wet, wet-to-dry, dry-to-wet, and dry-to-dry states can be altered, and homogeneity of the rainfall time series may be disturbed.

Replacement tests proposed by Teegavarapu (2012a) and illustrated in Figure 8-4 can be used to check the validity of radar data when missing rain-gauge values are replaced by radar-based rainfall values in three possible configurations: (1) systematic, (2) random, and (3) systematic and random replacement. In the case of systematic replacement, a specific portion (i.e., length of rain-gauge data) is replaced with radar data [area shaded dark in Figure 8-4(a)]. Similarly, random replacement involves replacing randomly selected rainfall observations with radar data [refer to Figure 8-4(b)]. A combination of the above two replacement procedures can also

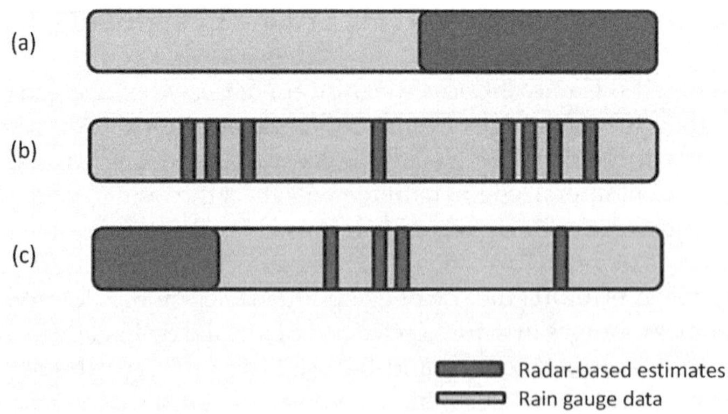

Figure 8-4. Data replacement tests with radar data replacing rain-gauge data: (a) systematic, (b) random, and (c) systematic and random.
Source: Teegavarapu (2015).

be used, as shown in Figure 8-4(c). Once the replacement of data is completed, the filled data sets can be tested for homogeneity using different homogeneity tests such as Pettitt's test (Pettitt 1979), Buishand's test (Buishand 1982), Alexandersson's standard normal homogeneity test (SNHT) (Alexandersson 1986), and Von Neumann ratio test (Von Neumann 1941). Statistical corrections of spatially interpolated precipitation data estimates in a recent study by Teegavarapu (2013) were effective in restoring site-specific statistics of the data with missing data filled.

Teegavarapu et al. (2009) and Teegavarapu (2012b, 2013) recommended the use of correction procedures based on SBE after initial spatial interpolation estimates are obtained. The correction procedures improve site-specific and regional statistics and also reduce the under- or over-estimation. Bias correction methods used in climate change studies can be used for correcting spatially interpolated rainfall data.

8.6 CONCLUSIONS

The focus of this chapter has been on spatial and temporal analysis of rainfall and estimation of missing gauge rainfall using radar-based rainfall estimates. Functional forms linking radar and rain-gauge data and GWO methods for estimating missing rainfall data were discussed in this chapter. Geospatial grid-based transformation of radar data from one spatial coordinate system to another was also briefly discussed.

REFERENCES

Alexandersson, H. 1986. "A homogeneity test applied to precipitation data." *J. Climatol.* 6, 661–675.

ASCE. 1996. *Hydrology handbook*. Reston, VA: ASCE.

ASCE. 2001a. "Artificial neural networks in hydrology. I. Preliminary concepts." *J. Hydrol. Eng.* 5(2), 115–123.

ASCE. 2001b. "Artificial neural networks in hydrology. II. Hydrologic applications." *J. Hydrol. Eng.* 5(2), 124–137.

Ashraf, M., J. C. Loftis, and K. G. Hubbard. 1997. "Application of geostatisticals to evaluate partial weather station network." *Agric. For. Meteor.* 84(3), 255–271.

Brimicombe, A. 2003. *GIS, environmental modeling and engineering*. Abington, Oxfordshire, UK: Taylor and Francis.

Buishand, T. A. 1982. "Some methods for testing the homogeneity of rainfall records." *J. Hydrol.* 58, 11–27.

Chang, K-T. 2006. *Introduction to geographic information systems*. New York: McGraw-Hill.

Chou, Y. H. 1997. *Exploring spatial analysis in geographic information systems*. New York: OnWord Press.

Cybenko, G. 1989. "Approximation by superpositions of a sigmoidal function." *Math. Control Signals Sys*. 2(4), 303–314.

Dingman, S. L. 2007. *Physical hydrology*. Long Grove, IL: Waveland Press.

Einfalt, T., K. Arnbjerg-Nielsen, C. Golz, N.-E. Jensen, M. Quirmbacj, G. Vaes, et al. 2004. "Towards a roadmap for use of RADAR rainfall data in urban drainage." *J. Hydrol*. 299, 186–202.

Fotheringham, A. S., C. Brunsdon, and M. Charlton. 2002. *Geographically weighted regression: The analysis of spatially varying relationships*. Chichester, UK: Wiley.

French, M. N., W. F. Krajewski, and R. R. Cuykendal. 1992. "Rainfall forecasting in space and time using a neural network." *J. Hydrol*. 137, 1–37.

Grayson, R., and G. Bloschl. 2001. *Spatial patterns in catchment hydrology: Observations and modeling*. Cambridge, UK: Cambridge Univ. Press.

Hodgson, M. E. 1989. "Searching methods for rapid grid interpolation." *Prof. Geograph*. 411, 51–61.

Hornik, K., M. Stinchcombe, and H. White. 1989. "Multilayer feedforward networks are universal approximators." *Neural Networks* 2(5), 359–366.

Hosking, J. R. M., and J. R. Wallis. 1998. *Regional frequency analysis: An approach based on L-moments*. Cambridge, UK: Cambridge Univ. Press.

Krajewski, W. F. 1987. "Co-kriging of radar and rain gage data." *J. Geophys. Res*. 92(8), 9571–9580.

Lloyd, C. D. 2007. *Local models for spatial analysis*. Boca Raton, FL: CRC Press.

Lloyd, C. D. 2010. *Spatial analysis data analysis: An introduction for GIS users*. Oxford, UK: Oxford Univ. Press.

Mather, P. M. 2004. *Computer processing of remotely-sensed images: An introduction*. New York: Wiley.

Navone, H. D., and H. A. Ceccatto. 1994. "Predicting Indian monsoon rainfall: A neural network approach." *Clim. Dyn*. 10, 305–312.

NOAA (National Oceanic and Atmospheric Administration). 2013. Accessed July 2013. http://www.nws.noaa.gov/oh/hrl/nwsrfs/users_manual/part2/_pdf/21hrapgrid.pdf.

O'Sullivan, D., and D. J. Unwin. 2010. *Geographical information analysis*. Hoboken, NJ: Wiley.

Pettitt, A. N. 1979. "A non-parametric approach to the change-point problem." *Appl. Stat*. 28(2),126–135.

Schuenemeyer, J. H., and L. J. Drew. 2010. *Statistics for earth and environmental scientists*. New York: Wiley.

Seo, D.-J., W. F. Krajewski, and D. S. Bowles. 1990a. "Stochastic interpolation of rainfall data from rain gages and radar using co-kriging—1: Design of experiments." *Water Resourc. Res*. 26(3), 469–477.

Seo, D.-J., W. F. Krajewski, and D. S. Bowles. 1990b. "Stochastic interpolation of rainfall data from rain gages and radar using co-kriging—2. Results." *Water Resourc. Res.* 26(5), 915–924.

Shepard, D. 1968. "A two-dimensional interpolation function for irregularly spaced data." *Proc., 23rd Nat. Conf. of the Assoc. for Computing Machinery*. New York: Association for Computing Machinery (ACM), 517–524.

Sibson, R. 1981. "A brief description of natural neighbor interpolation." *Interpreting multivariate data*, V. Barnett (ed.), 21–36. Hoboken, NJ: Wiley.

Simanton, J. R., and H. B. Osborn. 1980. "Reciprocal-distance estimate of point rainfall." *J. Hydraul. Div.* 106(7), 1242–1246.

Smith, M. D., M. F. Goodchild, and P. A. Longley. 2007. *Geospatial analysis: A comprehensive guide to principles, techniques and software tools*. Leicester, UK: Winchelsea Press.

Teegavarapu, R. S. V. 2007a. "Use of universal function approximation in variance dependent surface interpolation technique: An application in hydrology." *J. Hydrol.* 332:16–29.

Teegavarapu, R. S. V. 2007b. "Estimation of missing precipitation data using soft computing based spatial interpolation techniques." *AGU Fall Meeting Abstracts*. Washington, DC: American Geophysical Union.

Teegavarapu, R. S. V. 2012a. "Assessment and evaluation of NEXRAD-based rainfall estimates in South Florida Water Management District." *SFWMD Rep.*, West Palm Beach, FL: SFWMD.

Teegavarapu, R. S. V. 2012b. "Spatial interpolation using non-linear mathematical programming models for estimation of missing precipitation records." *Hydrolog. Sci. J.* 57(3),1–24.

Teegavarapu, R. S. V. 2012c. "Bias Assessment Tool (BAT), Florida Water Management District." *SFWMD Rep.*, West Palm Beach, FL: SFWMD.

Teegavarapu, R. S. V. 2013. "Statistical corrections of spatially interpolated precipitation estimates." *Hydrol. Proc.* 28(11), 3789–3808. DOI: 10.1002/hyp.9906.

Teegavarapu, R. S. V. 2014. "Missing precipitation data estimation using optimal proximity metric-based imputation, nearest-neighbor classification and cluster-based interpolation methods." *Hydrolog. Sci. J.* 59(11), 2009–2026.

Teegavarapu, R. S. V. 2015. "Data quality improvement for NEXRAD: Final report." South Florida Water Management District, October. West Palm Beach, FL: SFWMD.

Teegavarapu, R. S. V. 2016. "Spatial and temporal estimation and analysis of precipitation." In *Handbook of applied hydrology*. New York: McGraw Hill.

Teegavarapu, R. S. V., and V. Chandramouli. 2005. "Improved weighting methods, deterministic and stochastic data-driven models for estimation of missing precipitation records." *J. Hydrol.* 312, 191–206.

Teegavarapu, R. S. V., and P. Bajaj. 2008. "Optimal spatial weighting methods for estimation of missing rain gage records." *Proc., EWRI World*

Environmental and Water Resources Cong., R. W. Babcock and R. Walton (eds.). Reston, VA: ASCE.

Teegavarapu, R. S. V., and C. Pathak. 2008. "Infilling of rain gauge records using radar (NEXRAD) data: Influence of spatial and temporal variability of rainfall processes." *Proc., EWRI World Environmental and Water Resources Cong.*, R. W. Babcock and R. Walton (eds.). Reston: VA: ASCE. DOI: 10.1061/40976(316)406.

Teegavarapu, R. S. V., D. Peters, and C. Pathak. 2008. "Evaluation of functional forms of rain gauge–radar (NEXRAD) data relationships." *Proc., EWRI World Environmental and Water Resources Cong.*, R. W. Babcock and R. Walton (eds.), 1–9. Reston, VA: ASCE. DOI: 10.1061/40976(316)413.

Teegavarapu, R. S. V., M. Tufail, and L. Ormsbee. 2009. "Optimal functional forms for estimation of missing precipitation records." *J. Hydrol.* 374, 106–115.

Teegavarapu, R. S. V., C. Pathak, and S. Chinatalapudi. 2010. "Infilling missing precipitation data using NEXRAD data: use of optimal spatial interpolation and data-driven methods." *Proc., EWRI World Environmental and Water Resources Cong. 2010: Challenges of Change*. Reston, VA: ASCE. DOI: 10.1061/41114(371)473.

Teegavarapu, R. S. V., A. Nayak, and C. Pathak. 2011. "Assessment of long-term trends in extreme precipitation: implications of in-filled historical data and temporal window-based analysis." *Proc., EWRI World Environmental and Water Resources Cong*. Reston, VA: ASCE. DOI: 10.1061/41173(414)419.

Teegavarapu, R. S. V., T. Meskele, and C. Pathak. 2012. "Geo-spatial grid-based transformations of precipitation estimates using spatial interpolation methods." *Comput. Geosci.* 40, 28–39.

Teegavarapu, R. S. V., and A. Nayak. 2017. "Evaluation of long-term trends in extreme precipitation: Implications of infilled historical data and temporal-window based analysis." *J. Hydrol.* 550, 614–634.

Vieux, B. E. 2001. *Distributed hydrologic modeling using GIS*. Dordrecht, the Netherlands: Kluwer Academic, Water Science and Technology Library.

Von Neumann, J. 1941. "Distribution of the ratio of the mean square successive difference to the variance." *Ann. Math. Stat.* 12, 367–395.

Wallis, J. R., D. P. Lettenmaier, and E. F. Wood. 1991. "A daily hydroclimatological data set for the continental United States." *Water Resour. Res.* 27, 1657–1663.

Wei, T. C., and J. L. McGuinness. 1973. "Reciprocal distance squared method, a computer technique for estimating area precipitation." *Tech. Rep. ARS-NC-8*. North Central Region, OH: US Agricultural Research Service.

CHAPTER 9

ADVANCED TOPIC: DESIGN OF RAINFALL MONITORING NETWORKS

9.1 DESIGN OF RAINFALL MONITORING NETWORKS

A rainfall monitoring network refers to a network of gauging stations located in space or in a region used for measurement of rainfall. According to the World Meteorological Organization (WMO 2004), the analysis of rainfall data is much easier and more reliable if the same gauges and siting criteria are used throughout the networks. This should be a major consideration in designing precipitation measurement networks. However, these conditions are not applicable all the time, as the objectives of studies that design monitoring networks differ from region to region. The economic costs can be considerable, and the siting of rain gauges can be very difficult, especially in highly developed areas and cities.

9.2 RAIN-GAUGE NETWORK DENSITY

Rainfall monitoring network density is simply the number of rain gauges per area of a specific region. The recommendations of WMO (1994) for the minimum network density of a rainfall network for hydrometeorological purposes are given in Table 9-1. Dingman (2007) provides details of rain-gauge densities from different countries throughout the world to highlight the issue of lack of sufficient densities in many developing countries. The rain-gauge density is an important aspect of the monitoring network design. However, placement of rain gauges within a basin or region is critical for hydrologic monitoring and modeling purposes. In the context of radar data and bias corrections, the density of rain gauges needed to

Table 9-1. Minimum Density of Rainfall Gauges* in Different Regions

Type of region	Area (km^2) per gauge
Flat regions of temperate Mediterranean and tropical zones	600–900
Mountainous regions of temperate Mediterranean and tropical zones	100–250
Small mountainous islands with very irregular rainfalls	25
Arid and polar zones	1,500–10,000

*Nonrecording rain gauges.

directly estimate rainfall is much higher than that needed to perform good bias correction of radar rainfall data.

9.3 OPTIMAL RAIN-GAUGE MONITORING NETWORKS

Clustering and nonclustering of point patterns in space can be assessed using spatial statistical techniques. Considering rain gauges as points in space, these techniques can be extended to assess the rain-gauge monitoring networks. Spatial statistical tools provide several indexes to identify clusters and dispersed point patterns. Two widely known indexes are (1) the average nearest-neighbor distance tool and (2) G statistic [Getis–Ord statistic defined by Getis and Ord (1992)]. In the case of the average nearest-neighbor index, the distance between each rain gauge (i.e., point) and its nearest-neighbor's (rain gauge) location is calculated. The average distance of all nearest-neighbor distances is calculated. If this average distance is less than the average for a hypothetical random distribution, the distribution of the rain gauges is considered to be clustered. If the average distance is greater than a hypothetical random distribution of points (rain gauges), then the rain gauges are considered to be dispersed. The index is the ratio of the observed distance divided by the expected distance (expected distance is based on a hypothetical random distribution of the same number of rain gauges within an area). Index values greater than 1 are dispersed; those that are less than 1 are clustered. The general G-index is another indicator of cluster and dispersion of point features in an area. This index was defined by Getis and Ord (1992). The clustering of rain gauges is very common and is often caused by a lack of adequate planning and design of monitoring networks. In many situations, rain gauges are installed to fulfill project objectives driven by requirements of hydrologic and water quality modeling and management studies. In such situations, clustering is desirable, especially to assess the rainfall variability at a fine spatial scale.

9.4 OPTIMAL DENSITY AND MONITORING NETWORKS

Optimal density refers to the optimal number of rain gauges within a specific region. Few guidelines for the assessment of minimum density are described by WMO (1994) and Wright et al. (2013). This density is defined by first fixing an analysis block (a subregion of constant size in a region) and second by determining the optimal number of rain gauges in this block. The design of optimal monitoring networks requires a critical assessment of the information to be gained about the heterogeneity of the rainfall process between distances among the sampling locations. Fortin and Dale (2005) indicate that the design of an optimal spatial sampling scheme requires a careful balance between sampling locations that are too close to one another, thus not providing enough new information (data highly autocorrelated), and sampling locations that are too sparse, so that processes at other spatial scales introduce too much variability (Haining 1990).

9.5 OBJECTIVES FOR MONITORING NETWORK DESIGN

This section and the following sections describe rain-gauge monitoring network design based on variance and geostatistical methods. The geostatistical method of rain-gauge optimization is one of the methods that are available for network design. It is important to note that the design of a monitoring network is aimed at characterizing the spatial variability of the rainfall across a specific region. This generally should be the primary objective when only rain gauges are available in a region and a network of an optimal number of rain gauges is required. However, other secondary objectives related to placement of gauges can be due to other reasons such as (1) validation of modeling efforts, (2) rainfall measurements required to derive inputs for a hydrological simulation model, (3) requirement of a rainfall monitoring effort for a specific hydrologic or hydraulic project, and (4) adjusting radar rainfall data. The design of the network depends on several factors that include (1) placement of rain gauges to maximize the information obtained from the rain gauges, (2) existing network of rain gauges, (3) monetary cost involved in the placement (relocation and installation) of the rain gauges, and (4) feasibility of installation of a rain gauge at a location within a specific region.

9.6 OPTIMAL MONITORING NETWORK DESIGN

The number of rain gauges required in a specific region (e.g., watershed, basin) can be determined by conceptually simple methods based on the variance of rainfall in space or its variants. The variance of rainfall values is calculated based on the existing number of rain gauges in the region. Rakhecha and Singh (2009) provided a discussion of three methods based on the variance of rainfall. Rycroft (1949) used variance in space and allowable variance in the estimate of mean rainfall to determine the optimum number of rain gauges, which is given by Equation (9-1).

$$N = \left[\frac{2\upsilon}{\omega}\right]^2 \qquad (9\text{-}1)$$

where
N = optimum number of rain gauges,
u = variance, and
w = allowable variance in estimate of mean rainfall.

A method described by Rakhecha and Singh (2009) uses the mean monthly coefficient of variation (CV_m), the tolerance value of CV_m, τ, and the existing gauges, n, in Equation (9-2) to obtain the optimum number of rain gauges, N.

$$N = \left[\frac{CV_m}{\tau}\right]^2 n \qquad (9\text{-}2)$$

A minor variant of the above equation was proposed by Ahuja (1960) and is given by Equation (9-3).

$$N = \left[\frac{CV}{x}\right]^2 \qquad (9\text{-}3)$$

where x is the desired percentage error in the estimate of mean rainfall. All the above methods provide the optimum number of rain gauges, and this number is compared with the number of existing number of gauges. Additional rain gauges are added based on the predetermined density.

Geostatistics and spatial interpolation methods are combined in this illustrative example to obtain an optimal rainfall monitoring network. Details of this rainfall monitoring network design using radar data can be found in SFWMD (2006). To devise a monitoring network, long-term rainfall data is required. The rainfall data is initially analyzed for its variability across a region and regions of fixed areas (or analysis blocks) are established. This refers to stratification of the region into an array of analysis blocks of specific spatial resolution. Several methods of partitioning of spatial data exist, and they include k-means clustering, variance quad-tree, and others. The analysis blocks of prespecified sizes are further evaluated using geostatistical methods to model the variability of rainfall within each block through fitted semi-variograms for the rainfall data. The semi-variogram that is used to model the covariance structure of the rainfall data provides the de-correlation lengths as range parameters of the model. The standard error (SFWMD 2006) of mean is then used as a metric to identify the optimal number of ground rain gauges in this example. The standard error can be considered one measure of the accuracy of the monitoring network. The required number of rain gauges in a specific analysis block can be obtained by improving the accuracy of the mean rainfall measurement. The effective number of rain gauges was obtained from the correlation distance calculated from rainfall data and point variance from gauge data for each analysis block. Haan (2002) and Matalas and Langbein (1962) described the information contained in a number of correlated stations. The information contained in data from n stations in a region having a correlation of ρ is equivalent to the information contained in n' uncorrelated stations (rain gauges) in the region, where n' is given by Equation (9-4).

$$n' = \frac{n}{1 + \rho_{ave}(n-1)} \tag{9-4}$$

where
n′ = effective number of stations,
n = total number of stations, and
ρave = average CC among CC stations.

Equation (9-4) suggests that as n increases, n' approaches $1/\rho_{ave}$. The effective number of rain gauges n' was related to separation distance "h." The covariance structure defined by spatial CC can be modeled by exponential semi-variogram or any empirical semi-variogram that can be fitted to rainfall data in each analysis block. The relationship between the spatial CC, ρ_{ave}, the separation distance, h, and the range parameter, R, is given by Equation (9-5).

$$\rho_{ave} = e^{\left(-\frac{3h}{R}\right)} \tag{9-5}$$

where σ^2 is the point variance and n' is again the effective number of stations. The standard error of the mean (SE) was calculated for all the blocks using Equation (9-6).

$$SE = \sqrt{\frac{\sigma^2}{n'}} \tag{9-6}$$

A scaled standard error (SFWMD 2006) SE' is defined in Equation (9-7).

$$SE' = \frac{\sqrt{\frac{\sigma^2}{n'}}}{SE_{n'=1}} \tag{9-7}$$

The standard error is calculated based on effective number of rain gauges (n'). Depending on the value of the average CC, range parameter and inter-rain gauge distance, the values of n' will tend to increase up to a threshold value and then decrease as the number of rain gauges increases.

9.6 OPTIMAL NETWORK DESIGN USING RADAR DATA

The optimum network design is based on the standard error, and the number of gauges in each spatial grid of fixed tessellation is referred to as the analysis block. Based on the covariance structure of the rainfall data, uniform or nonuniform analysis block sizes may be identified. SFWMD (2006) outlined two methods of optimum network design in a study related to rain-gauge network optimization. The network designed is optimal considering the stipulations imposed on the block size and other constraints. However, it should be noted that a strict mathematical programming model (i.e., an optimization model) is not used in the network design example presented in the next section.

9.6.1 Variable Density Analysis Block Approach

In this approach, the point variance (σ^2) for each block is used to identify the number of rain gauges (n) that have a CC (ρ) to achieve a desired level of constant accuracy defined by the magnitude of the standard error (SE). The relationship is given by Equation (9-8).

$$n_k(SE)^2 = 1 + (\rho_k(n_k - 1)\sigma_k^2) \quad \forall k \tag{9-8}$$

where k identifies each analysis block. Once the SE value was fixed, the above equation was solved for n_k. This equation can be solved for different values of SE for each of the analysis blocks using any optimization solver. The value obtained for n needs to be rounded off to the nearest integer as the number of rain gauges. The approach considers only the number of rain gauges within a block. Further study would be needed (using land-cover data, etc.) to assess the exact placements of rain gauges within a given block. Uniform standard error can be adopted for all the analysis blocks, and the optimum number of rain gauges can be obtained for each block. Once the optimum number is obtained for each block, this number is compared with the existing number of rain gauges in each analysis block. If the existing

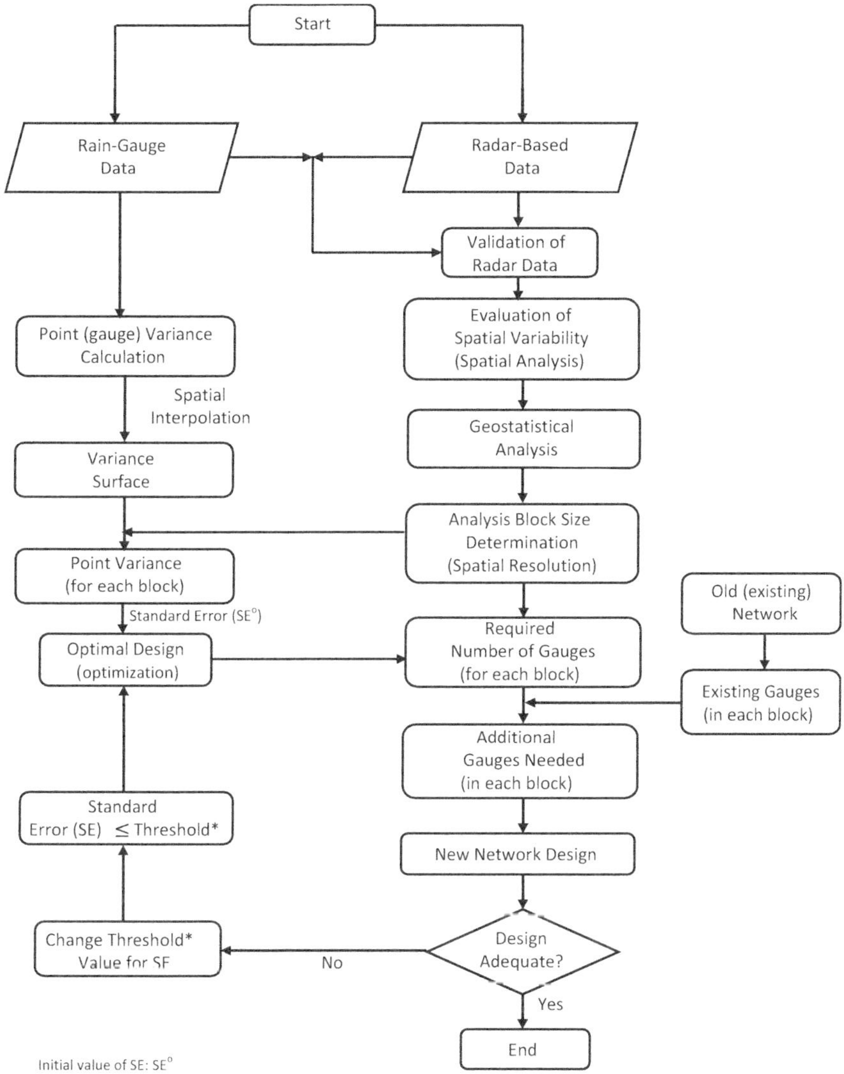

Figure 9-1. Flowchart of steps utilized in optimal design of a rain-gauge network.

number of gauges in a block is higher than the optimum number, then a recommendation can be made to remove the additional number of gauges. When the existing number of gauges is less than the optimum, additional gauges can be recommended for that block. A flowchart describing the complete set of steps for design of monitoring network is shown in Figure 9-1. Steps shown in the flowchart are similar to those used in a recent study for the design of a solar radiation sensor network (Teegavarapu et al. 2015). The optimal design of a network depends on the spatial variability of rainfall, the analysis block size defined based on region-specific characteristics, and the standard error used as a metric of network estimation accuracy.

9.7 POST-NETWORK DESIGN RECOMMENDATIONS FOR RAIN-GAUGE PLACEMENTS

The placement of rain gauges should be based on the locations of existing and proposed rain gauges and the following considerations:

- Proposed rain gauges to be added may include both new and spatially relocated rain gauges.
- Relocation of rain gauges within a specific analysis block needs to be undertaken considering the inter-rain–gauge placement distance.
- If an analysis block already has the required number of rain gauges, relocation should only be considered within the block when the spacing between the rain gauges is lower than the optimal separation distance.
- Cost structure (installation, maintenance, etc.) needs to be incorporated into decision making regarding the installation of new and relocation of existing rain gauges within and across the analysis blocks.
- Excess rain gauges from one or more analysis blocks can be utilized in analysis blocks where the number of required rain gauges is larger the existing number.
- Locations of the rain-gauge placements in analysis blocks should be based on hydrometeorological considerations.

A separate study needs to be carried out to define the implementation strategies for the recommended network design. A cost-based assessment of the different implementation strategies is recommended as the next step in this process. Also, a feasibility study of using optimization models in a multiobjective framework that maximizes the information obtained from the rain gauges and minimizes the cost structure for installation and maintenance needs to be evaluated.

9.8 IDENTIFICATION OF METEOROLOGICALLY HOMOGENEOUS AREAS

Meteorologically homogeneous rainfall areas can be developed using radar data. An example of homogenous rainfall areas developed by Pathak et al. (2009) for the southeastern part of the state of Florida is shown in Figure 9-2. Radar-based rainfall data sets at a 4 km × 4 km (2.49 mi × 2.49 mi)

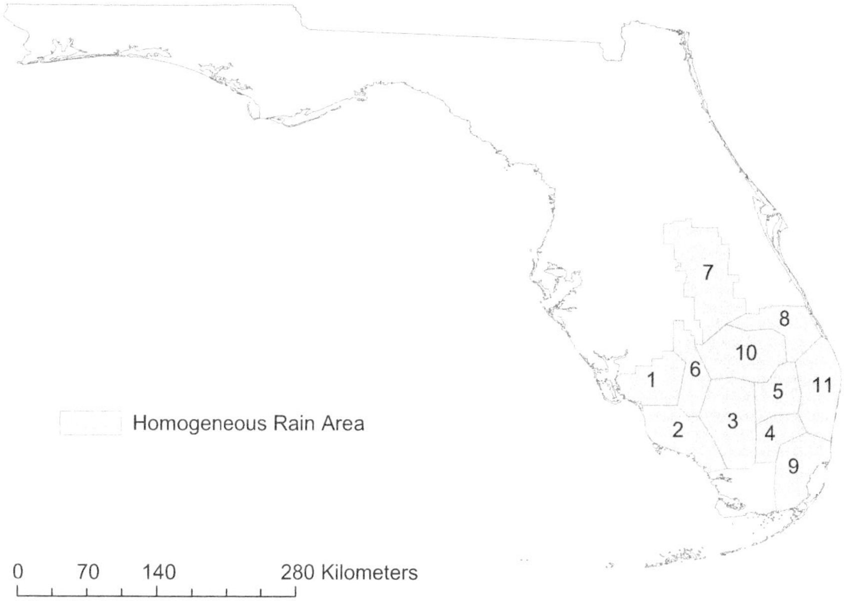

Figure 9-2. Meteorologically homogeneous rainfall areas identified using radar data in South Florida.

grid spatial resolution were initially used to define regions with similar rainfall characteristics and links to spatially varying meteorological phenomena (e.g., sea breeze). These regions are referred to as "seed areas." Correlation mapping by evaluating correlations of rainfall accumulations in these seed areas to rainfall amounts in all the 4 km × 4 km (2.49 mi × 2.49 mi) grids to identify and define regions having homogeneous rainfall in space and time was then carried out. These homogeneous areas were used as units for evaluation of spatial and temporal variability of precipitation extremes in a study by Teegavarapu et al. (2013).

9.9 CONCLUSIONS

The chapter presented a summary of precipitation measurement network density and design. The main focus was on the use of rain-gauge and radar-based rainfall estimates for the design of optimal precipitation monitoring networks. A framework that used a geostatistical approach for the development of rainfall monitoring networks was presented. Use of radar data for defining meteorologically homogeneous rainfall areas was briefly discussed in this chapter.

REFERENCES

Ahuja, P. R. 1960. "Planning for precipitation network for water resources development in India." *WMO Flood Control Series, Report No. 15*, 106–112.

Dingman, S. L. 2007. *Physical hydrology*. Long Grove, IL: Waveland Press.

Fortin, M. J., and M. Dale. 2005. *Spatial analysis: A guide for ecologists*. Cambridge, UK: Cambridge Univ. Press.

Getis, A. and J. K. Ord. 1992. "The analysis of spatial association by use of distance statistics." *Geograph. Anal.* 24, 189–206.

Haan, C. T. 2002. *Statistical methods in hydrology*. Ames: Iowa Press.

Haining, R. P. 1990. *Spatial data analysis in the social and environmental sciences*. Cambridge, UK: Cambridge Univ. Press.

Matalas, N. C., and W. B. Langbein. 1962. "Information content of the mean." *J. Geophys. Res.* 67(9), 3441–3448.

Pathak, C. S., M. Onderlinde, and H. E. Fuelberg. 2009. "Use of NEXRAD rainfall data to develop climatologically homogeneous rain areas for central and south Florida." *Proc., EWRI World Environmental and Water Resources Cong.*, Reston, VA: ASCE, 1–11. DOI:10.1061/41036(342)618.

Rakhecha, P. R., and V. P. Singh. 2009. *Applied hydrometeorology*. New Delhi: Springer.

Rycroft, H. B. 1949. "Random sampling of rainfall." *J. S. Afr. For. Assoc.* 18, 71–81.

SFWMD (South Florida Water Management District). 2006. "Rain gauge network optimization study: Final report." West Palm Beach, FL: Vieux & Associates.

Teegavarapu, R. S.V., A. Goly, and J. Obeysekera. 2013. "Influences of Atlantic multidecadal oscillation phases on spatial and temporal variability of regional precipitation extremes." *J. Hydrol.* 495, 74–93.

Teegavarapu, R. S. V., C. S. Pathak, J. R. Mecikalski, and J. Srikishen. 2015. "Optimal solar radiation sensor network design using spatial and geostatistical analyses." *J. Spatial Sci.* Accessed August 2018. http://dx.doi.org/10.1080/14498596.2015.1051147.

WMO (World Meteorological Organization). 1994. *Guide to hydrological practices*. Geneva: World Meteorological Organization.

WMO. 2004. "Handbook on CLIMAT and CLIMAT TEMP reporting." *WMO/TD-No. 1188*. Geneva: World Meteorological Organization.

Wright, D. B., J. A. Smith, G. Villarini, and M. L. Baeck. 2013. "Long-term, high-resolution, radar rainfall records for urban hydrology." *J. Am. Water Resour. Assoc.* 50(3), 713–734.

INDEX

aggregation, 5
artificial neural networks (ANNs), 97, 100–101

backscatter, 33–35
beam blockage, 17, 19, 19*f*
below-beam effects, 17, 18*f*
bias analysis methods: background of, 73–74; categories of, 76–78, 77*f*; detailed description of, 77–84, 78*t*, 79*f*–84*f*; ideal performance measures and skill scores and, 84, 85*t*, 86*t*; indexes based on metrics and, 75–76; rain gauges and, 76; types of, 75; utility assessment indexes and performance measures and, 85, 87*t*, 88–89
bias assessment tool, 79, 79*f*
bias corrections: background on, 89; with limited rain-gauge data, 90; temporal resolution issues and, 90–91
bright bands, 18

Climate Data Online (CDO), 7, 8
climatology-calibrated precipitation analysis, 47–48
correlation weighting techniques, 97

data replacement tests, 106–107, 106*f*
depth area adjustment factors (DARFs), 28
depth area reduction factors, 66
deterministic weighting, 95–96; missing data estimation and, 96–98
digital hybrid reflexivity field (DHR), 40
digital precipitation array (DPA), 40, 42
digital storm total precipitation (DSP), 40, 42
distributed models, 6–7
Doppler radar, 4–5
dual polarization, 27

elevated radar sites, 17
equal weights method, 105
error distribution, of radar rainfall estimates, 40–41

gauge-adjusted radar rainfall estimation (GARR), 20–21, 27–28, 60–62
gauge–radar adjustment: approaches to, 45–47, 46*f*; local bias correction, 43; mean-field bias correction, 41–42; quality control and, 73
gauge–radar observation merging, 43–45
gauge–radar pair (GRP), 42
geographically weighted optimization (GWO), 101–102, 102*f*
geospatial grid-based transformations, 102–105, 104*f*
gridded multi-radar products, 48–49

ground clutter, 17, 19
groundwater recharge modeling, 68

Hydrologic Rainfall Analysis Project grid (HRAP grid), 14–15, 15*f*, 23
HEC-METVUE (Meteorological Visualization Utility Engine), 21–22
Hurricane Dennis (2005), 45, 46*f*
hybrid scan discontinuity, 18
hydrologic model evaluation, 75
hydrologic modeling: calibration of, 60–61; data requirements for, 59–60, 60*f*; radar rainfall data for, 60–62
hydrometeor identification, 37–38

infilling rainfall data: issues related to, 105–107, 106*f*; radar data use and, 98–102, 101*f*, 102*f*
inverse distance weighting method (IDWM), 3, 96, 97

Journal of Hydrologic Engineering, 1–2

kriging technique, 3

local bias correction, 43
Louisville Jefferson County Metropolitan Sewer District (Louisville, KY), 24

maximum area method, 104
mean-field bias (MFB), 41–42; evaluation of, 73, 77
meteorologically homogeneous rainfall areas, 118–119, 119*f*
multisensor precipitation estimator data (MPE data), 7–8, 23
multi-sensor products, 48–49

National Centers for Environmental Information (NCEI), 7
National Weather Service (NWS), 8, 23, 27
native radar data resolution, 11–12, 12*f*
NCDC Weather and Climate Toolkit, 11–12
near-real-time flood warning system, 67
NEXRAD program, 3–4, 8, 15, 26
NOAA Central Library of US Daily Weather Maps, 8
NOAA (National Oceanic and Atmospheric Administration), 3, 27

optimal density, 113
optimal *Z-R* relationships, 54–57
overhead "blind" spots, 18

phased array system, 27
precipitation observations, access to, 48–49
PRISM climate group, 7
PRISM Climate Group, 8
probable maximum precipitation (PMP) estimation, 24–25

quality control, gauge-adjustment procedures and, 73
quantile–quantile plot (Q–Q plot), 82
quantitative precipitation estimates (QPE): daily precipitation reports and, 47–48; function of, 31, 45, 47 (*See also* radar rainfall data processing)

radar beam issues, 17, 17*f*, 18*f*, 19, 19*f*
radar coverage limit discontinuity, 18, 19
radar data: anomalies in, 17–19, 18*f*–20*f*; optimal network design using, 116–118, 117*f*; used for infilling rainfall data, 98–102, 101*f*, 102*f*
radar data acquisition (RDA) subsystem, 33, 34*f*
radar precipitation products, 33
radar product generator (RPG), 33, 34*f*, 41
radar quality control, 35, 36*f*, 37, 40
radar rainfall data: archived, 27; availability in United States of, 7–8, 8*f*; data formats and resolutions of, 13–16, 15*f*, 16*f*; dual polarization and, 27; geospatial grid-based transformations of, 102–105, 104*f*; hydrologic modeling and, 59–62;

native resolution of, 11–12, 12*f*; optimal *Z-R* relationships and, 54–57; phased array and, 27; QA/QC and data management and, 16–20, 17*f*–21*f*; radar mosaics and, 12–13, 13*f*, 14*f*; rainfall measurement methods and, 53; rain-gauge and radar based measurements and, 53–54; scientific formats for gridding, 13–14

radar rainfall data analysis: depth area reduction factors and, 66; function of, 21–22; rainfall frequency analysis and, 65–66

Radar Rainfall Data and Application Task Committee [Watershed Council of the Environmental and Water Resources Institute (EWRI)], 2

radar rainfall data applications, 66; groundwater recharge modeling, 68; near-real-time flood warning system, 67; rain-gauge network design, 68–69; sewer system modeling, 67–68

radar rainfall data evaluation/improvement: optimal *Z-R* relationships and, 54–57; rainfall measurement methods and, 53; rain-gauge and radar based measurements and, 53–54

radar rainfall data processing: access to precipitation observations and estimates and, 48–49; applicability of gauge radar approaches and, 45–47; background related to, 31–32; daily precipitation reports used with radar QPE and, 47–48; error distribution of estimates and, 40–41; gauge-radar adjustment and, 41–43, 41*e*, 42*e*; gauge-radar observation merging and, 43–45, 46*f*; hydrometeor identification and, 37–38; radar QPE products and, 40; radar quality control and, 35, 36*f*, 37; reflectivity-precipitation rate relationships and, 38–39, 39*f*; single-polarization radar precipitation estimates and, 33–35, 34*f*; stages in, 33

radar rainfall data use: missing data from rain gauges and, 26; probable maximum precipitation estimation as, 24–25; real-time hydrologic monitoring, flood forecasting and disaster management as, 23–24, 23*f*–25*f*

radar rainfall estimates: background of, 3–4; conclusions related to, 27–28; critical issues and limitations associated with, 88–89; derived from conversion of reflexivity, 73; error distribution of, 40–41; function of, 4–7; gauge-adjusted, 20–21; overview of, 1–2

rainfall data: missing data estimation and, 96–98; spatial and temporal analysis of, 95–96

rainfall estimation: background of, 1; radar, 1–7; references for, 8; traditional, 2–3, 3*f*, 4*f*

rainfall frequency analysis, 65–66

rainfall interpolation techniques, 3, 4*f*

rainfall monitoring networks: description of, 111; meteorological homogeneous areas identification and, 118–119, 119*f*; network design and, 113; optimal density and monitoring networks and, 113; optimal network design and, 114–116; optimal rain-gauge monitoring networks and, 112–113; post-network design recommendations and, 118; radar data and, 116–118, 117*f*; rain-gauge network density and, 111–112, 112*t*

rainfall transformation process, 59, 60*f*

rain-gauge networks. *See also* rainfall monitoring networks: density of, 111–112, 112*t*; design of, 68–69

rain-gauges. *See also* rainfall monitoring networks: measurement of, 2–3, 3*f*; missing data estimation and, 26; placement of, 118; reports on, 49

range fade, 19

real-time hydrologic monitoring, flood forecasting and disaster management, 23–24, 23*f*–25*f*
reflectivity-precipitation rate relationships, 38–39, 39*f*
reflectivity-to-rain rate (*Z-R*) equations, 38
replacement tests, 106–107, 106*f*
Rice University and Texas Medical Center (Houston, TX), 24
River Forecast Center (RFC) (NWS), 7, 8*f*

sewer system modeling, 67–68
single-polarization radar precipitation estimates, 33–35, 34*f*, 37, 38
South Florida Water Management District (SFWMD), 24, 25*f*, 65, 68, 69, 114, 115; radar rainfall estimation in, 63–65
spatial analysis: issues related to, 95–96; missing data estimation and, 96–98
spatial domain filters, 103
spatial weighting method, 103
standard hydrologic grid (SHG), 14–16, 16*f*

Taylor's diagram, 80, 80*f*, 81*f*
temporal analysis, issues related to, 95–96
Terminal Doppler Weather Radar (TDWR), 31
Thiessen polygon technique, 3, 4*f*, 97

University Corporation for Atmospheric Research (UCAR), 48–49
USACE HEC, 21–22
US Army Corps of Engineers (USACE), 14, 15, 23, 27
utility assessment indexes, 85, 87*t*, 88–89

Vance Air Force Base (AFB) radar, 26
variable density analysis block approach, 116–117, 117*f*

Weather Service Radar Doppler (WSR-88D), 31–33, 34*f*, 61